HANDBOOK OF MAINTENANCE MANAGEMENT

HANDBOOK OF MAINTENANCE MANAGEMENT

Cost-Effective Practices

Second Edition

by

Frank Herbaty

Frank Herbaty & Associates
Maintenance Management Consultants
Bolingbrook, Illinois

np

NOYES PUBLICATIONS
Park Ridge, New Jersey, U.S.A.

Library of Congress Catalog Card Number: 89-8696
ISBN: 0-8155-1204-X
Printed in the United States

Published in the United States of America by
Noyes Publications
Mill Road, Park Ridge, New Jersey 07656

10 9 8 7 6 5 4 3 2 1

Library of Congress Cataloging-in-Publication Data

Herbaty, Frank.
Handbook of maintenance management.

Rev. ed. of: Cost-effective maintenance management. c1983.
Bibliography: p.
1. Plant maintenance. 2. Industrial equipment--Maintenance and repair. I. Herbaty, Frank. Cost-effective maintenance management. II. Title.
TS192.H47 1989 658.2'02 89-8696
ISBN 0-8155-1204-X

This book is dedicated to my beloved wife, Linda, who taught me all the important things about life, none of which are taught in school.

Preface

I have had many gratifying phone calls and letters complimenting me on the first edition of this book. The most common statement was that they could tell without any question that I had hands on maintenance experience, an attribute they found missing from many of the other books they read. Many said they read the book several times and made it mandatory reading for all supervisors.

This second edition was written principally to incorporate the comments and suggestions received from the many readers of the first edition. I am indebted to all who took the time out to send me these critiques. In addition I obtained a wealth of information for improving the text as a result of giving seminars based on the text.

Two suggestions predominated. This included a suggestion to program the computer systems I describe in the text and the other is that I should write a section on Fleet Management. I have accomplished both of these tasks.

Chapter 23 entitled "Accounting and Computer Systems" has been completely rewritten and renamed "Computer Systems." This chapter contains my philosophy on the use of the computer in maintenance and a description of my computer programs.

The major addition to the book is Part 5, "FLEET MANAGEMENT" which includes the management of the diverse moving equipment items that are normally a part of any industrial plant, i.e., cars, trucks, material handling equipment, road maintenance equipment and fire trucks, etc. It varies in importance with each company, and is a very important function of utility companies and municipal governments. Most industrial plants have fleet type equipment and therefore must be concerned about this function. Although many of the maintenance management techniques are the same as for any other plant equipment, there is enough difference to warrant separate treatment.

Finally, many new case studies and figures have been added and a list of these has been added to help organize what is now a much larger book.

I know that these additions will make the book more useful and I hope that suggestions for improvement continue to arrive. In this regard I would like to quote from the Preface to the first edition:

> "Since those early days my maintenance knowledge has been constantly growing. Every time I think I know it all, I learn more. Every time I become certain of how to do something, I learn a better way. Will it ever end? I don't think so."

Bolingbrook, Illinois
September 1989

Frank Herbaty

Preface to First Edition

My love affair with maintenance started when I graduated from college in 1944. I joined the Merchant Marine and worked in the engine room. The chief engineer soon found out that I didn't know the difference between a monkey wrench and a Stillson wrench. This, coupled with the fact that I was a graduate mechanical engineer, earned me his undying contempt and frequent beatings with his fists when I did or said anything to displease him. I might add that he was about 6'6", 260 pounds and built like Hercules! In those days, before 100% unionization, the captain and chief engineer literally had a life and death control over the crew. If you made them mad, watch out!

I learned fast, and rose to the rank of chief engineer in two and one half years, something impossible to do during non-war years, when it might normally take at least ten years. Back in those days, each engineer had to know how to operate and maintain everything, including mechanical, electrical, electronic, structural, radio, and all other types of equipment aboard ship. After a year, my pulse quickened every time I walked down into the engine room. There were the characteristic smells, sounds and vibrations: steam, oil, heat, water, pumps, motors, boilers, evaporators. I got to the point where I could tell by listening and by smelling if everything was O.K. by the instantaneous integration of the hundreds of different sounds and smells. I was hooked, I was in love with maintenance.

Since those early days my maintenance knowledge has been constantly growing. Everytime I think that I know it all, I learn more.

Everytime I become certain of how to do something, I learn a better way. Will it ever end? I don't think so. How long does it take to design, implement and analyze the worth of a preventive maintenance program? Ten years? And how does *your* system compare with someone else's system? There really isn't enough time in life to find out. Besides, Plant Engineers don't stay long enough with a company to find out. The person who has the best chance of finding the best way is the maintenance management consultant. He has the opportunity of seeing and evaluating many different systems and is, therefore, in the best position to pick a winner.

It is important that you know something about my background so that you will know where I am coming from and that you will be convinced that I have the credentials to tell you that *I have found the answer to optimization of the maintenance management function.* That is what this book is all about. This is my background.

I have been a maintenance management consultant and have held many other jobs, all involved with maintenance. The variety of jobs has given me a view of maintenance from every possible angle and has helped me in my understanding of what is, and is not important. My experience covers consulting engineering (10 years), marine engineering (5 years), plant engineering, maintenance and production (16 years), management consulting (4 years) and research (5 years), total experience 40 years.

In these categories I have held many high level positions including Vice President of Pope, Evans and Robbins, a well known consulting engineering firm, Plant Engineering Manager for Rohr Corp., Director of Engineering, Operations and Maintenance at Kennedy Space Center, Director of Research for President Carter's Reorganization Project, Senior Program Manager at the Department of Energy for three large research programs and was cited by a leading trade journal as being one of the top five maintenance consultants in the country.

In giving lectures and seminars throughout the world, I have found the telling of anecdotes from my past experience to be very effective in illustrating the point being discussed. For this reason, I have made liberal use of this technique as case studies in the text. I have also decided to use the informal approach in writing and will, therefore, write as if I am speaking to you in my living room. I feel confident that the use of the optimization principles expounded in this text will bring dramatic improvement to any maintenance organization.

One more point. I don't intend to philosophize about alternate ways of doing things. I've paid my dues and made my choice and have selected the best system for each function to the best of my knowledge. As a result, only one system is described for each function. Although the system described may not exactly fit every plant, it can easily be modified to suit any plant. I have had no problem doing so on any of my assignments. My use of one system has its advantages since I go into sufficient detail for the reader to implement the system.

This handbook is divided into four parts. A brief explanation of the content of each of these parts follows:

Part 1

This part covers the approach to maintenance management optimization. It defines what maintenance is and how its cost is measured and states that effort must be applied in proportion to the impact the effort has on:

- Company goals and objectives (or mission);
- Profit, cost effectiveness and life cycle cost;
- Safety;
- Environmental impacts; and,
- Product quality (or mission quality).

Parts 2 and 3

Parts 2 and 3 discuss downtime and productivity control. These are two out of possibly twenty subjects that may be discussed under the subject of maintenance management. I have found, beyond a shadow of a doubt, that the importance of these two subjects totally dwarfs the impact the other subjects have on the maintenance management function.

Part 4

Recognizing that much has been written on maintenance support functions, such as material, engineering, purchasing, computers, personnel and the union, etc., I confine my remarks in Part 4 to things that I have learned that are unique or helpful in managing these support functions.

There are a few books in existence on maintenance management. They include all functions that are legitimately a part of the subject. This is the first book however, that emphasizes optimization and differentiates between the trivial and the two most important functions of a maintenance manager by presenting a practical system for optimizing maintenance and concentrating corrective effort where it will reap the highest rewards.

Bolingbrook, Illinois
June 1983

Frank Herbaty

Introduction

I can sum up my experience as a maintenance management consultant by using with slight alterations Winston Churchill's famous World War II statement:

> "Nowhere in the history of the maintenance profession throughout the USA has the work of so many been fouled up royally by so few."

Interpreted this means, sadly, that management is usually at fault if the maintenance unit is malfunctioning, not the hourly workers. As a matter of fact, most of the information I have gotten on what the problem is comes from the hourly workers. I have found three common serious problems, these are:

(1) Lack of a cooperative spirit between management and labor to do what is necessary to accomplish the company's mission, or a lack of incentive to do so. The incentive is needed for all, not just the executives as is the case in some companies. Man frequently uses only a small fraction of his capabilities; these must be unlocked by incentives. Practically all of the problems I have resolved in my consulting assignments were the result of simple common sense principles.

(2) Management failure to supply workers with fundamental things necessary to get the job done in a timely manner. In this regard, managers must feel that it is absolutely necessary for suggestions to come from everyone, not just supervisors in order for every one to feel that they have a part in the company's success. I have found too many companies where the philosophy is that hourly workers are

bordering on being mentally incompetent and that you can not expect any good ideas from them. As a matter of fact management frequently attempted to prevent me from talking to the hourly workers. This tragic philosophy is aptly illustrated in case study 3C on page 187.

(3) Another problem I have experienced lately, particularly with the advent of computers, is the use of complex management systems that tend to confuse and lower productivity rather than increase it, or conversely the lack of any system at all. In the first case, management gets tied up with the overly complex system rather than with getting the job done and in the latter case, management has no system for resolving complex problems with the result that the problem perpetuates for years. The moral to this experience is that systems must be designed to fit the need by someone who knows what he is doing rather than, as is frequently the case, by a computer expert or accountant type with no maintenance background, who designs the system to give you "anything you could ever possibly want." I have taken great pains to tailor the system to the need, with minimum complexity, in my computer systems. However, the most important point to understand is that the lack of a computer system is never the main problem.

Other common problems that I have found are:

- No optimization systems
- No criticality coding systems
- Inadequate or non-existing preventive maintenance system
- No failure analysis system
- Inadequate or non-existent repair records
- No system for apportioning labor and material expenditures with criticality
- Inadequate control of some factors affecting productivity coupled with ignoring most of the factors

As I have stated previously, all of these problems can be resolved with the copious use of common sense. This book will make it easier for you to use common sense by sticking to basic, useful principles as opposed to scholarly, theoretical concepts that are interesting, but will never be used in practice.

About the Author

Frank Herbaty, with over 40 years experience in the maintenance management field, is presently president of Frank Herbaty Associates, Inc., of Bolingbrook, Illinois, a maintenance management consulting firm. He is also a registered professional engineer in New York, Arizona, and Florida. He has a long and distinguished career in maintenance management which includes being Senior Program Manager and Director of Engineering, Construction, and Facilities Management for the U.S. Department of Energy; Director of Research and Special Projects for the President's Reorganization Project; and Director of Engineering, Operations, and Maintenance at the Kennedy Space Center from 1964 to 1971. Previously, he was also Vice President and Chief Engineer for Pope, Evans, and Robbins of New York City, a well known consulting engineering firm.

In addition, Mr. Herbaty has written numerous articles, presented papers, given lectures and taught classes on Maintenance Management at various conventions, seminars, colleges, public and private sector organizations throughout the world. Mr. Herbaty was president of the Florida Chapter of AIPE as well as Group Vice President, and Chairman of the International Public Relations Committee of AIPE. He has also been listed by a leading trade magazine as one of the nation's top five maintenance consultants.

NOTICE

To the best of our knowledge the information in this publication is accurate; however the Publisher does not assume any responsibility for the accuracy or completeness of or consequences arising from, such information. Mention of trade names or commercial products does not constitute endorsement or recommendation for use by the Publisher.

The book is intended for informational purposes only. Final determination of the suitability of any information or product for use contemplated by any user, and the manner of that use, is the sole responsibility of the user. Expert advice should be obtained when implementation is being considered.

Contents and Subject Index

PART 3
PRODUCTIVITY CONTROL

PART 4
MAINTENANCE SUPPORT SYSTEMS

PART 5
FLEET MANAGEMENT

List of Figures

List of Case Studies

Part 1

Maintenance Optimization Approach

1

General

If a maintenance manager had all of the following resources he would have no need for the maintenance optimization systems described in this book:

- o Ten times the labor force he has presently.
- o All highly skilled workers.
- o All the material, spares, tools and equipment he needs to do the job in endless supply.
- o No need to worry about costs.
- o Downtime whenever he wants it, to work on facilities and equipment.

However, in the real world, for all practical purposes, none of the above is normally available, therefore common sense dictates that the available supply of labor, materials, tools and equipment be used wisely and where they will do the most good in accomplishing the companies' mission. The art of doing this is a skill I have called: *"MAINTENANCE OPTIMIZATION."*

Why then have I found, in my experience, that practically no companies are using all these optimization systems? Instead of me answering this question I invite you instead to find out what your company is doing in this regard and to determine if the optimization systems make sense to you and are worthy of being implemented. I assure you that *I know* they are effective beyond a shadow of doubt.

Maintenance is performed to help achieve the goals and objectives of a company or organization. In the case of a manufacturing organization, the objective is to produce and sell a product at a profit. In many organizations, the objective is simply to perform a mission in a cost-effective manner. For example, a mission of the strategic air command is to keep the required number of bombers in the air at all times in a cost-effective manner. The maintenance optimization principles described in this text apply equally as well to any type of organization, be it product or mission oriented, (the latter being the case in most government organizations).

Optimization of maintenance is the art of performing maintenance, within a set of constraints, in a manner that best achieves the goal of the company. One set of constraints might be to state the sum of maintenance labor and material, *plus* production loss costs caused by inadequate maintenance, must be optimized to the minimum value. In addition, other things must also be considered, for example, safety and environmental impact. If all factors are considered in the art of optimization, it becomes very complex.

The optimization approach about to be described has the following characteristics:

- There are many optimization systems, not one. Each approaches the problem in a slightly different manner.
- Most all of the optimization systems are cyclical in nature and are designed to lessen the problem each time a cycle is complete until finally, there is no more problem.
- The optimization systems recognize that the primary task of the maintenance department is to perform maintenance work and therefore the manhours for analysis are hard to come by and must

not be squandered on nonproductive effort. For this reason a very important part of the optimization system is to establish methods for determining criticality of both effort and facilities and equipment.

- Finally, downtime control (Part 2) and productivity control (Part 3) are of such importance in optimizing maintenance, that the entire emphasis in this text is placed on these two subjects. This is done because the other maintenance subjects either have trivial impact or they are normally not a major problem.

With these thoughts in mind several of the most important general optimization systems are described, keeping in mind that downtime and productivity control systems (Parts 2 and 3) are optimization system also. It should also be noted that one can enter into each cycle at any point, since the cycle repeats itself many times. This text, however, is written from the viewpoint of an existing organization with existing goals and objectives, rather than from the viewpoint of a plant starting up for the first time.

The optimization systems are:

- Auditing Priorities
- Management Cycle
- Maintenance Goal
- Optimizing Preventive versus Breakdown Maintenance
- The "Magic Tool"
- Criticality Coding
- Maintenance by Industry Type
- Tailoring PM to need
- Downtime Improvement Cycle
- Productive Control

2

Audit Priorities

Every manager should perform periodic audits to ensure that existing goals, objectives, policies and procedures are being properly executed. It is recommended that the following auditing priority system be followed to ensure that analytical effort be applied in the proper sequence and to insure that this effort is not squandered on the wrong function. (For example, there would be no point in auditing job methods when a crew is spending 30% on travel time, 30% material acquisition time, 10% obtaining tools, 20% idle and only 10% of the time working. In this case, audits should be conducted on travel and material acquisition times first).

PRIORITY 1 - CONTROL COMPETENCE OF MAINTENANCE LABOR

The first priority is to insure the availability of technically competent mechanics who can repair anything that breaks down. This is the most obvious to the owner of a small plant who is about to hire his first maintenance man. There is no question in *his* mind that this is the first priority. Control is achieved by:

- o Careful hiring practices
- o Training
- o Motivation

PRIORITY 2 - CONTROL OVER THE "LOGIC" OF WORK ASSIGNMENTS

This item can best be illustrated by giving a few examples of illogical work assignments that should be caught and stopped by the audit, i.e.,

Construction Work: It is illogical to have maintenance men build a desk for $300, when; a superior desk can be purchased for $200.

Janitorial Work: It is illogical to have an electrician clean offices when the work should be performed by a lower paid janitor.

Design Changes: It is illogical to make design changes or modifications by unqualified operation or maintenance personnel. Checks and balances must be instituted to insure that authorized and qualified personnel work on appropriate work.

Maintenance Work: It is illogical to perform maintenance work that can be eliminated by correcting a design error (replacing a fuse that is constantly blowing with one of a correct size).

It is illogical to perform maintenance work that can be eliminated by correcting an operational practice that causes damage to facilities (craneman damaging structural elements by collision).

It is illogical to perform repetitive maintenance tasks that can be eliminated by a revision to preventive maintenance procedures. It is important to realize that some repetitive maintenance procedures can become so ingrained over the course of many years that the illogicalness of the procedure is overlooked.

PRIORITY 3 - CONTROL OF PRODUCTION LOSS CAUSED BY EQUIPMENT DOWNTIME

The cost of production loss caused by equipment downtime must be minimized by the optimization techniques described in Part 2.

PRIORITY 4 - CONTROL OVER LABOR PRODUCTIVITY

Then, systems must be designed to obtain high labor productivity as described in Part 3.

PRIORITY 5 - MANAGEMENT INFORMATION AND CONTROL SYSTEMS

Finally, management information and control systems must be installed to "fine tune" the maintenance function.

PROBLEMS IN AN EXISTING ORGANIZATION SHOULD BE CORRECTED IN THE ABOVE ORDER OF PRIORITY!

3

"Management" As an Optimization System

The functions performed by management, constitute an optimization system of the greatest importance. The functions are cyclical in nature and are repeated over and over again, never ending, each time, improving the total performance of the organization. The functions of management are:

- Establish goals and objectives, policies and procedures.
- Establish permissible variance from these guidelines.
- Measure performance to established guidelines.
- Compare performance measurement feedback information to guidelines.
- Isolate and identify deviations beyond tolerances.
- Determine basic cause for deviations.
- Determine corrective action.
- Plan method of implementing corrective action.
- Schedule plan for implementing corrective action.
- Implement corrective action.

o Follow-up to ensure completion of corrective action and to prevent overswing.

These functions are shown graphically in Figure 1-1.

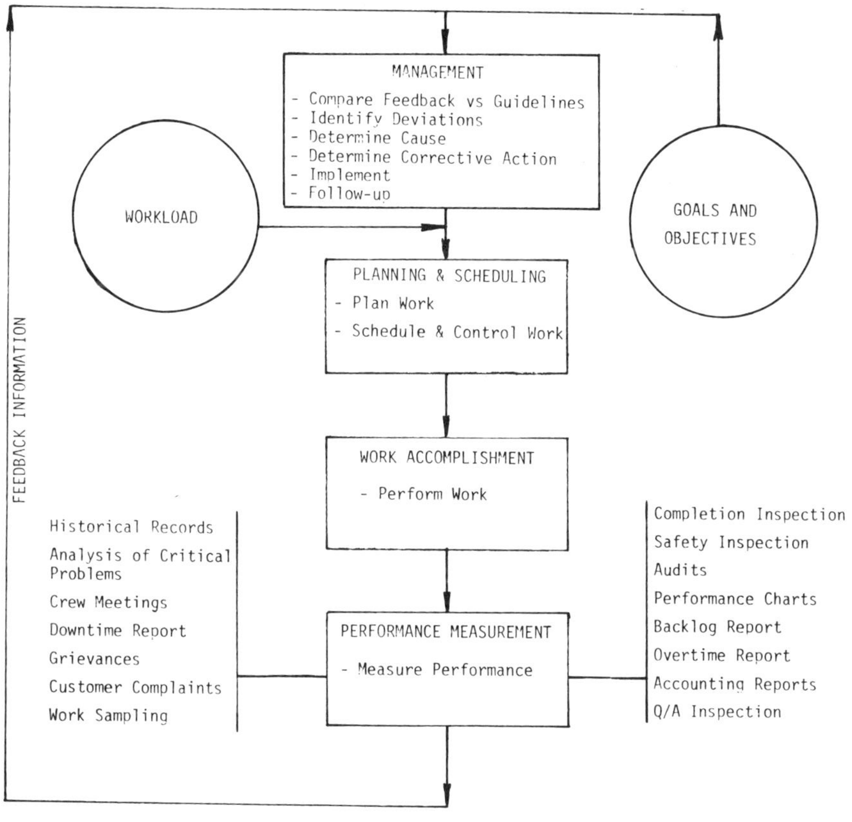

Figure 1-1: Functions of management.

Stated more briefly, management must establish goals and objectives then design a performance measurement system to measure performance to these goals and objectives. The goals and objectives and the policies and procedures written to implement them may be a long and complex list of guidelines. Just writing the guidelines will not insure that they are implemented. They may not be implemented for one or all of the following reasons.

- o Employees don't understand the guideline.
- o Employees don't like the guideline.
- o So many procedures have been written that the employee doesn't have time to carry out all of them.
- o Employee likes his way better.
- o Etc.

It is important to understand that the management cycle is universal and applies to everything even, for example, the factors affecting productivity.

Most of the companies I have serviced have no written policies, goals, or objectives for their maintenance department with the result that everyone is going off in different directions. This reminds me of the famous saying, "The lame leadeth the blind and both falleth into the ditch." I have observed directly the destructive effects of companies that set up no procedures for doing anything. On the other hand many government agencies have so many procedures that when an employee wants to do something, he must find out how the regulation tells him to proceed rather than to think through the solution himself. This emasculates the employee and has a tendency to destroy:

- o Innovation
- o Reasoning power
- o Initiative
- o Inventiveness (because it takes a superhuman effort to change anything to the extent that no one has the energy to try)
- o Lowers productivity

In short, a system for writing a procedure for practically everything lowers productivity and dulls the thinking process.

Obviously, the point to be at is somewhere in between having no procedures and having too many. It takes a rational mind and common sense to determine the optimum point.

Managers must perform periodic audits of all guidelines and study feedback reports, otherwise employees will be doing what they want to do, which may be nothing. Another good time to audit procedures is during a major crisis, breakdown or accident, since violation of established guidelines is frequently the cause. A few case studies by the author will illustrate the point.

CASE STUDY 1A: The Frozen Safety Valve

As was my practice at company X, I periodically performed random audits of procedures. I was auditing the functions of a material expeditor. I closed my eyes and reached into the drawer and took out a purchase requisition at random from the expeditor's desk. It was marked "Urgent Safety Item" and was three weeks old. It was an order for a new safety valve for the cafeteria boiler. Investigation revealed that a mechanic, while performing a check by lifting the safety valve leavers, had found it frozen. He immediately reported this to his foreman; however, nothing had been done about the problem in three weeks. I immediately ordered the boiler to be shutdown and the valve overhauled; a task which was completed in four hours. At least four company procedures were violated. Both of these employees were severely disciplined for this serious infraction of company rules. I had previously led an investigation of a boiler explosion in a cafeteria which killed dozens of people because the engineer forgot to remove boiler safety valve gags. The reason this job went sour is because the foreman failed to separate and perform his tasks by criticality. Even though the job was labeled "Urgent Safety Item," the foreman was handling the jobs by the fifo method (first in, first out).

CASE STUDY 1B: The Burning of a Welder

A welder had just completed burning out a section of floor plate inside an electrical cubicle. When he was finished, he lifted up the cut-out section and let it fall against high voltage bus-bars. This caused an explosion and severe burning of the welder by flying pieces of molten copper. Investigation of ths incident revealed that the following company procedures had been violated:

- o Unauthorized entry into a cubicle sealed by Q/A.
- o Maintenance performed during a launch countdown without the launch director's written approval.
- o The foreman had failed to brief his crew on the safety hazards of the job and failed to initial the box on the work order certifying that he had done this.
- o The welder was not supposed to be in the cubicle without being with a journeyman electrician.
- o The power was not shut off per regulations.
- o Written approval for adding a new line to the cubicle had not been obtained.
- o The job was started by the 8 to 4 shift foreman who failed to brief the 4 to 12 shift foreman on the job status.

There was a violation of my Project Manager concept which states that every project must have *one* and only one Project Manager assigned to it and that his responsibility is to worry about *all* tasks of the project excluding nothing. He should have seen to it that none of these violations happened. When a project goes wrong it is the Project Manager's responsibility to conduct an investigation such as illustrated in the case study above, then take steps to insure that it does not happen again. These subjects are discussed in complete detail in Chapters 11 and 16.

The maintenance manager must realize that one reason for non-compliance to an established procedure, policy goal or objective could be that the employee has no time left to do his job if he were to carry out all written procedures. He will then resort to "greasing the squeaky wheel" and only performing those procedures that are expedited. The manager must counteract this tendency to write too many procedures by approving only those procedures that optimize direct labor effort.

In summary, it must be emphasized that the management control cycle is a continuous, never ending optimization process, which requires constant surveillance on the part of the manager. In order for it to work, every department must have established goals, objectives, policies, procedures and a matching and well thought out performance measurement system.

4

Maintenance Goal

The goal of any well run maintenance organization is to have the lowest cost of the sum of two quantities, i.e:

- o Maintenance labor and material; and
- o Production loss cost resulting from an inadequate maintenance program.

The achieving of the lowest cost is an optimization technique and is shown graphically in Figure 1-2. Notice that when maintenance cost is zero (no maintenance) the production loss cost is naturally at its highest. As maintenance effort is intelligently increased, the production loss gradually decreases until the lowest combined cost is achieved. This is the maintenance goal. Maintenance effort applied beyond this point, *increases* cost.

This phenomenon is true of any service organization, for example, purchasing and stores. Its costs are also the sum of its labor and material costs plus the losses caused by inadequate service to its client.

Unfortunately, this fact is not often recognized or accepted in industry as is illustrated by the following case studies.

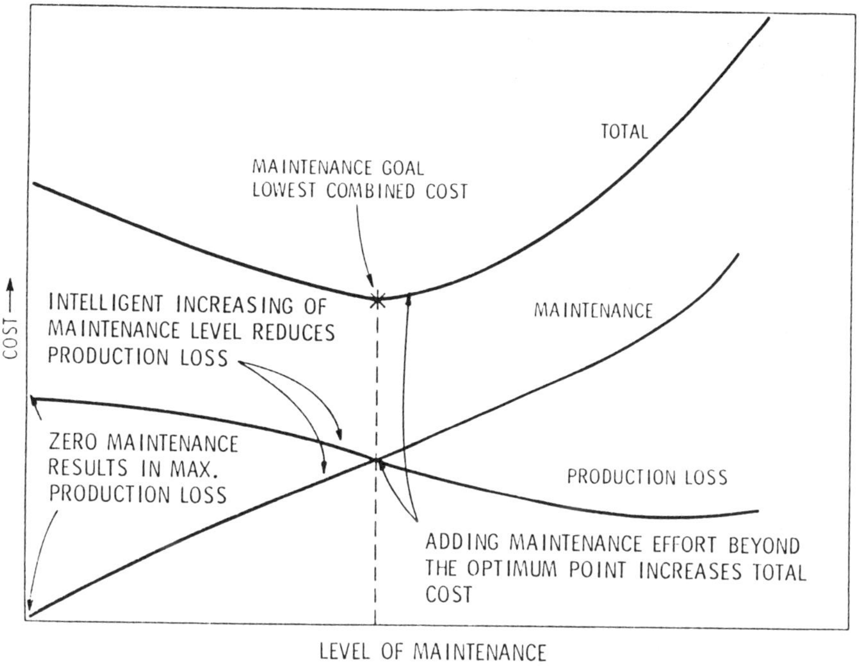

Figure 1-2: Maintenance goal.

CASE STUDY 1C: Who Should Be Charged?

While doing a fleet management study, I noticed that the downtime on all construction vehicles used to put in new electric and gas services and transmission lines was excessive. When a crew of thirteen workers was dispatched to a job, a critical vehicle frequently broke down (such as a bucket truck or pole hole driller). The crew would usually wait for someone to pick them up and tow the vehicle back to the shop. This normally wasted the entire day for the crew while waiting for their vehicle to be fixed. The lost time was charged to the construction work order.

I argued that this method of charging time did not put pressure where corrective effort was required: on the fleet manager. I won the battle and thereafter lost time caused by vehicle breakdowns was charged to the fleet manager, who was then raked over the coals at weekly meetings to reduce this lost time cost. He got the message and an improving trend was soon noticed.

CASE STUDY 1D: The Efficient Purchasing Manager?

At a certain plant, I was doing a study of the engineering, maintenance, purchasing and stores departments. A quick survey of the maintenance department personnel quickly surfaced material acquisition as their biggest problem. This was not surprising since the mine and plant were in an isolated, remote area. Excessive stock outages quickly led me to the purchasing department.

The purchasing manager was indignant about being audited and was proud of the fact that he had just made a study of his department which resulted in his laying off ten out of the 20 people who worked for him. This was done at the prodding of top management who were anxious to reduce indirect costs.

I asked him if he made any attempt to determine the impact his layoff had on the 5,000 people he served. He hadn't, of course, and gruffly mentioned that this was none of his business and that his job description called for him to run an efficient *purchasing* department.

He was not sensitive to the concept shown in Figure 1-2. Needless to say, his concept of "purchasing efficiency" changed shortly since his department's performance was the major factor inhibiting adequate maintenance performance.

Figure 1-3 shows the two control systems used to control the two costs shown in Figure 1-2, i.e.

Cost	Control System
Maintenance labor and material	Productivity
Production loss	Downtime

The intelligent application of these two control systems has a greater influence on the success of a maintenance organization than any other factor.

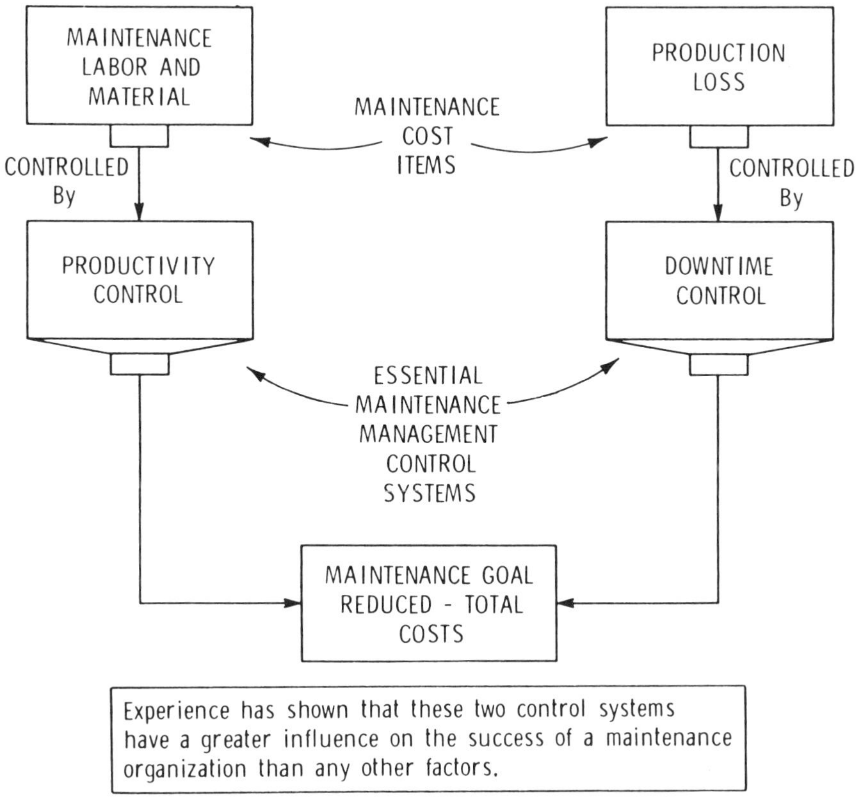

Figure 1-3: The two essential control systems.

5

Criticality Determination

GENERAL

Obviously not all equipment, materials, functions, and labor efforts are of equal importance relative to the mission, or goals and objectives of a particular organization. This fact *should* lead all of us to the following conclusions:

- It is of the utmost importance that everyone in the company understand fully the mission, or goals and objectives of the company.
- That equipment, materials, functions and labor effort should be categorized by criticality to the company's mission, goals and objectives, so that emphasis can be placed on those items having the highest criticality.

Although the author has been in the business for many years, he has yet to find a company that does an effective job at the two tasks listed above. Even worse, most companies don't even make the attempt. Presented here are several methods of categorizing by criticality and examples that illustrate how the methods are used, what happens when you don't do this and company employees are not familiar with the company's objectives. (In general, the impact is a squandering of the company's valuable labor and material resources and low productivity!)

THE "MAGIC TOOL"

The "Magic Tool" is a concept that recognizes that normally a large number of jobs account for a very small percent of the total labor hours or cost and conversely, a very small number of the jobs normally account for a very large percent of the total labor hours and cost. The improvements that can be made with this "Magic Tool" can be so dramatic that I have purposely named it the "Magic Tool" to draw attention to it for like many other things the concept is nothing new but for the most part has been stored away in the attic, unused by most maintenance organizations!

This concept is true of most functions in an organization! Knowledge of this fact is of tremendous value in making criticality determinations. The relationship can be surfaced by writing a very simple computer program that sorts information already in the computer to facilitate criticality determinations. The computer program that the author markets has this capability. See Chapter 23 and Appendix. Figure 1-4 illustrates the technique. Notice that in this figure, only one-half (0.5) of 1% of the number of equipment items out of 1,000 account for 77% of the annual maintenance cost. Obviously, these five equipment items should receive the lion's share of maintenance effort if cost alone is used as a criteria. Figure 1-5 shows the magic tool concept graphically. The following case studies illustrate the use of the magic tool.

Item Number	Machine Number	Description	Annual Maintenance Cost Dollars	Annual Maintenance Cost Percent	Cumulative Cost Percent	Cumulative Number of Items Percent
1	001	-	xxx	30	30	0.1
2	008	-	xxx	29	59	0.2
3	109	-	xxx	10	69	0.3
4	110	-	xxx	5	74	0.4
5	005	-	xxx	3	77	0.5
6	854	-	xxx	1	78	0.6
7	562	-	xxx	1	79	0.7
8	687	-	xxx	0.4	79.4	0.8
9	295	-	xxx	0.3	79.7	0.9
10	346	-	xxx	0.3	80.0	1.0
↓	ETC	-	ETC	ETC	ETC	ETC
↓	ETC	-	ETC	ETC	ETC	ETC
1000	↓		↓	↓	↓	↓
TOTALS			xxx	100	100	100

Figure 1-4: A "Magic Tool" spreadsheet.

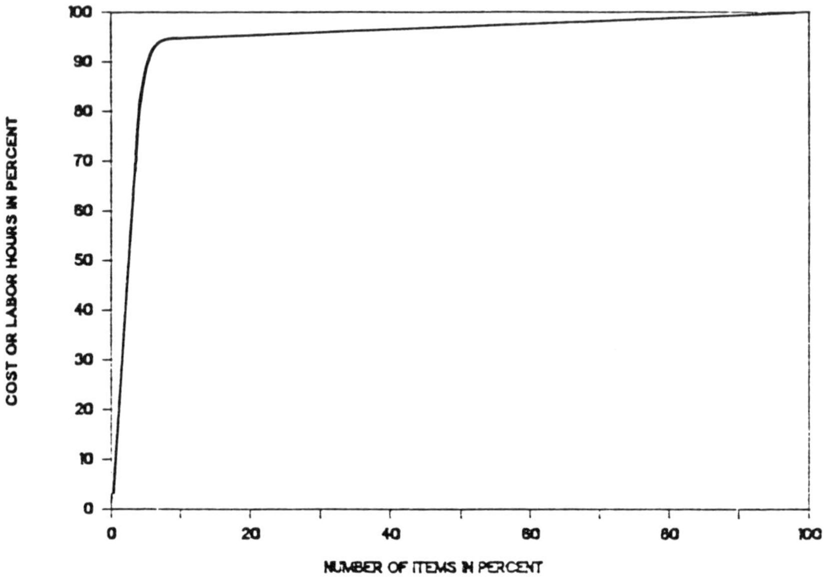

Figure 1-5: The "Magic Tool" shown graphically.

CASE STUDY 1E: A Statistical Analysis

A chart identical to that shown in Figure 1-4 was made for a process plant being studied. The chart revealed the following:

- o 1% of 10,000 equipment items accounted for 80% of the annual maintenance cost.
- o 50% of the 10,000 items accounted for only one-tenth (0.1) of 1% of the annual costs.

Further study revealed that the preventive maintenance (PM) program was applied uniformly over all 10,000 items. The following action was taken:

- o 50% of the items were removed from the computer and preventive maintenance programs and put on a 100% breakdown maintenance program, except for those items requiring lubrication. This greatly simplified the computer run analysis and permitted emphasis to be given to the more important equipment.

- The remaining 50% of the items were coded by criticality representing decreasing mission criticality, and the codes entered into the computer program as part of the equipment number. A program was initiated to tailor the thoroughness of the preventive maintenance program to the criticality of the equipment.
- It was decided that 100% of the maintenance analysis effort would be put immediately on the 1% of the items accounting for 80% of the cost.
- It was decided to make a chart similar to Figure 1-4 for all 30,000 line items of stock and another chart for downtime and production loss cost versus equipment item. (Parts usage is a barometer or equipment failure.)

The new charts revealed the following:

- Ten items of stock out of 30,000 accounted for 70% of the annual stock replacement cost.
- 20,000 items of stock accounted for only 1% of the annual cost.
- 200 out of 10,000 equipment items accounted for 85% of the annual downtime hours.
- Only two out of the 10,000 equipment items accounted for 100% of the production loss cost!

The following action was taken as a result of the new charts:

- A special task force was assigned to investigate the ten stock items accounting for 70% of the cost with the objective of either solving the wear or failure problem, obtaining lower prices or otherwise, reducing costs. These ten items were found to coincide with the 1% of the equipment items causing 80% of the maintenance cost.

10,000 items were eliminated from stock and 500 items were transferred to open bench stock.

- A special system was designed by myself for investigating the startling revelation that only two equipment items accounted for 100% of the production loss cost. (See part 2, "The Principle of Simultaneous Downtime.") It is interesting to note that downtime does not necessarily cause a production loss. Reasons for downtime not causing a production loss may be:

  - redundant equipment available
  - equipment is producing into in-process storage
  - stockpiles in the distribution system
  - plant was not operating 24 hours per day, 7 days per week and therefore, loss could be made up.

The many reductions in equipment and stock items previously on the weekly computer run made the run dramatically smaller so that the maintenance analysts had a fighting chance to read it each week. Furthermore, the coding by criticality and placement in descending order by cost made it possible to concentrate only on those items that were critical by virtue of cost, downtime or criticality. As a minimum, the magic tool should be used on:

- Annual equipment maintenance cost
- Stock replenishment cost
- Downtime and/or production lost cost

CASE STUDY 1F: The Crusher Balls

An analysis of spare parts costs at a plant was made similar to that shown in Figure 1-4. This chart revealed that crusher balls (4" diameter steel balls) were one of the three highest cost items and cost approximately $5,000,000 per year to replace. These balls were used in 25, 1,000 HP ore crushers. The balls were purchased from six suppliers. It was decided to run a test by dividing the crushers into six groups and then using one supplier for each of the six groups. Cost of balls per ton of throughput was measured over a period of one month. Each ball varied slightly in material composition and manufacturing technique.

The results of the test showed that one ball cost 20% less per ton of throughput. Orders for others were cancelled and the optimum ball was used exclusively thereby, saving $1,000,000 per year.

CASE STUDY 1G: What Is Your Mission?

Most government agencies have their missions described in a public law passed by Congress. Some of the larger agencies have multiple and complex missions. Although the missions are clear in the law and in the top echelons of management, it is not uncommon for the worker three or four levels down in the organization structure to completely lose sight of the agency's mission. The following example will illustrate the point:

It was observed that one very large government agency, which had a maintenance mission, had several hundred forms with accompanying procedures, which had to be filled out as part of performing the various maintenance functions. In interviewing the supervisor in charge of forms and procedures and asking him what his mission was, he stated that his mission was to generate forms and procedures as stated in his job description. This is a common fault in many job descriptions, which fail to even mention or make clear what the mission of the organization is.

It was explained to the supervisor that his function was indirect staff support to the direct mission, which was maintenance. Indirect support can only be justified by means of either improving the productivity of direct labor or by fulfilling a mandatory function such as safety. For example, when a foreman is hired to supervise fifteen maintenance people, it is done in anticipation of his increasing work productivity by say, 20%, thereby saving three workers for the price of one supervisor. In a similar manner, it was explained that it was the supervisor's job to write forms and procedures only to the extent that they increase maintenance labor productivity. It was quite obvious to anyone knowledgeable about maintenance management that the issuance of several hundred forms and procedures would actually *decrease* productivity soon after the number of forms exceed ten or twenty. In a similar manner, the safety engineer may think that safety is the company's mission, the procurement specialist thinks that it's his job to uphold the integrity of the procurement system, and the personnel man, the integrity of the personnel system, etc., etc.

RECOMMENDATIONS FOR MAGIC TOOL USE

Plant personnel in the case studies thought they had a good feel for the criticality of the various maintenance factors involved, but were surprised at the actual statistics when plotted as previously described. This lack of specific knowledge of plant criticality factors is more often the case than not. This approach can and must be used in *all* areas of maintenance management if effective use of limited personnel is to be made.

Suggestions for other systems to be analyzed in this manner are given in the following tabulation:

- o Number of downtime occurrences versus equipment items
- o Maintenance cost per ton or per square foot or per dollar value of facility versus department
- o Stock parts cost versus specific part
- o Dollars production loss versus equipment item
- o Job type versus cost

- o Analysis of repetitive failures by equipment item, failure type, cost per failure and failure frequency
- o Numbers of equipment items versus criticality code
- o Productivity percentages versus maintenance area
- o Productivity percentages versus maintenance trade
- o Productivity percentages versus foreman
- o Productivity percentages versus job type
- o Union grievances by type, foreman and trade

CRITICALITY CODING

Each plant has many thousands of parts, machines, people, and problems. If trivia are not separated from critical items, the system will collapse under its own mass of data, paperwork, and instructions.

Therefore, it is very important to code equipment by criticality. The code should be added to the equipment number to facilitate sorting by the computer. Only after criticality is determined should maintenance efforts be applied. Several factors should be considered in determining criticality:

Dollar Value: A $500,000 piece of equipment is more important than a $100 piece of equipment and should receive greater maintenance care.

Function: Equipment vital to production should receive greater maintenance attention, particularly if downtime will result directly in lost production.

Constant Maintenance Problems: Equipment that frequently breaks down should receive greater attention.

Safety: Failure of certain equipment may constitute a threat to life.

Other factors that affect criticality will depend on problems relating to a specific plant's needs, and the criticality coding system should be tailored to the particular facility. Example of two simple coding systems follow:

Code A - Critical production equipment for which there is no backup or substitute.

Code B - Critical production equipment with backup or in-process storage.

Code C - Noncritical equipment.

or

Code 1 - Critical equipment--downtime may result in loss of sales.

Code 2 - Semicritical equipment--downtime may result in lost production that can usually be recovered.

Code 3 - Noncritical equipment--downtime seldom affects production.

Once the coding system is designed, maintenance efforts should be apportioned according to criticality. Such effort will be in the form of:

- o Breakdown maintenance labor and materials
- o Preventive maintenance labor and materials
- o Design of maintenance optimization systems
- o Maintenance engineering analysis

If breakdowns on ten of 10,000 equipment items are causing 99% of the production loss, a major portion of the maintenance analysis effort should be spent on these

10 items. Also, the Preventive Maintenance program on these items should be strengthened accordingly. If you are applying your maintenance analysis effort uniformly across all 10,000 items of equipment, then you are not using common sense!

DIRECT VERSUS INDIRECT LABOR

A very useful concept used by practically all companies is the concept of direct versus indirect labor measurement and ratios. Direct labor is labor working directly on the mission of a company. For example, if a company is producing chairs, anyone directly working on chairs is a direct worker. Anyone indirectly supporting chair production tion is indirect, for example, a janitor. Companies set goals for desirable ratios, for example, it is common to consider 20 indirect to 80 direct workers as a reasonable ratio. Furthermore, companies keep track of the ratios of their competition, because if the ratio of the competition is superior, than their company may not be competitive.

In a sense, this concept is somewhat similar to the concept of criticality coding, in that we don't want to go overboard in the number of indirect workers we hire just as we don't want to spend too much time on non-critical equipment.

The following tabulation is useful in getting our thinking straight concerning where maintenance fits in:

- o Direct labor works directly on company's mission.
- o Indirect labor supports or increases productivity of direct labor.
- o All maintenance is indirect.
- o All systems, procedures, computer systems paperwork are indirect and are justified only if they improve direct labor performance, (assuming that none of the above are the mission of the company).

If we were to leave the definition as described above

(as it would be by the controller of the company) then the concept of direct versus indirect labor ratios would be useless in the maintenance department because everything would be indirect.

However, if the maintenance manager describes his mission as maintenance, then he can use the concept by dividing his department into direct and indirect according to that definition. For example the following would be a reasonable definition of direct and indirect labor in the maintenance department:

Direct Labor	Indirect Labor
Electricians	Manager
Plumbers	Foreman
Mechanics	Supervisor
Welders	Planners
Etc.	Estimators
	Etc.

A reasonable ratio of direct to indirect would be 15 to 1. Similarly, materials can be labeled direct and indirect, i.e.:

Direct Materials	Indirect Materials
Lumber	Toilet paper
Oil and grease	Computer paper
Nuts and bolts	Pens and pencils
Etc.	Etc.

Similarly tools and equipment can be labeled:

Direct Tools and Equipment	Indirect Tools and Equipment
Machine tools	Air conditioning equip.
Wrenches & screwdrivers	Computers
Ladders	Etc.
Etc.	

It is important to keep track of these ratios to make certain that you are not going overboard in your procurement of indirect items because taken as a whole indirect items only pay off it they improve the productivity of the direct labor as discussed for example in case study 1G.

6

Type of Industry

The percentage that maintenance cost is of the total operating budget varies over a very wide range. For example, a service organization may rent office space where the rental cost includes maintenance. This organization would have no maintenance people on their payroll.

Many manufacturing organizations have maintenance costs that vary between 5-10%.

On the other hand, many process plants have maintenance costs that vary between 50-95% and the maintenance percentage of these and other types of plants is increasing due to automation and robotization.

The maintenance manager should be aware of the maintenance percentage for several reasons:

- He must be aware of the percentage of his competitors. If his cost is higher, the lower cost company may run his company out of business.
- He must be aware of the availability percentage of critical equipment, lest competitors run him out of business with higher percentages. Examples of critical equipment that requires and usually gets detailed attention as far as tracking the productive operating time percentages are:

- Utility boilers
- Cement kilns
- Paper machines

The reason for this, of course, is that downtime on this equipment can result in loss of millions of dollars in revenues. This is discussed in greater detail in Chapter 12, "Principle of Simultaneous Downtime."

- o If the maintenance percentage is running around 1%, none of top management will have much interest in sophisticated maintenance systems. If on the other hand, the maintenance percentage is about 95%, use of the latest maintenance techniques will be warranted and top management will be scrutinizing his activity, perhaps on a daily basis, and if he does a good job, he may eventually become general manager of the plant. The maintenance manager, with costs at 1%, will have much less of a chance in achieving a higher position.

In summary, the maintenance manager must be aware of the maintenance percentage for two reasons:

1. To determine the level of sophistication in maintenance system design.
2. To have a feel for the relative importance of his division to the total manufacturing activity.

7

Preventive Maintenance (PM) Policy

GENERAL

The determination of which equipment items should be included in the preventive maintenance program and the degree of preventive maintenance is part of the maintenance optimization program. The methods of determining criticality previously discussed should be the basis for making this determination. Lack of an intelligent analysis of these two factors will increase maintenance costs. The large number of companies that decide to include *all* equipment when starting a preventive maintenance program for the first time is astonishing!

All members of a task force charged with designing and implementing a new preventive maintenance program should establish a policy for the design of the PM system which incorporates the following policy factors:

- All effort expended on designing a new preventive maintenance program is an expense subtracting from profit.
- Preventive maintenance tasks will *increase* maintenance costs when first initiated, until the beneficial effect of the preventive maintenance task has time to take effect because at initiation you have both the PM and the breakdown costs.

- The degree of PM should be related to criticality of equipment, and the type of PM (as described in the next few pages) should be selected to suit the criticality and particular operating problems of the equipment.

- A preventive maintenance task may *permanently* increase costs, if the author of the task doesn't know enough about maintenance. (This point is made because some organizations assign the function to unknowledgeable or clerical personnel under the mistaken theory that all they have to do is copy the instruction from the manufacturer's maintenance manuals.)

- Production loss cost should be optimized with maintenance cost as described in Chapter 4 with the realization that all breakdowns do not cause a production loss but usually cause a loss in productivity.

- Preventive maintenance should be optimized with breakdown maintenance as described in Chapter 13.

- Finally, at the start of the preventive maintenance program, people must be selected to both design and implement the program. Normally these people already have jobs and there may be pressure, in some cases, to assign the least competent personnel available. This tendency must be vigorously combated, particularly in the case of people assigned to design the system, who should be the *most competent* people in the organization.

With these points in mind, management should proceed as follows:

- Assign the most competent people to design the system.
- Start with the most critical equipment (determined from a computer printout by criticality code), and schedule the instruction writing and implementation tasks in accordance with the manpower available, incrementally, so as not to overtax the available manpower.
- The implementation rate should be designed to give time for the preventive maintenance instruction to reap its benefit in reduced labor, material and production loss costs. Ideally, the preventive maintenance task labor added should equal the savings from previously installed preventive maintenance. In this manner, the only additional effort required is for the first small group of preventive maintenance tasks implemented. In other words, preventive maintenance labor is substituted for breakdown maintenance labor on an incremental and equal basis. By the time the entire preventive maintenance is implemented, a net savings in total maintenance cost should be realized.
- Production loss should be optimized against maintenance costs as described in Chapter 4.
- Preventive maintenance instructions should be optimized as described in Chapter 12.

These concepts are so important that I have summarized them in the two tables shown below:

PM POLICY FACTORS

- Relate degree of PM to equipment criticality.
- PM causes increased costs initially

- o Poorly conceived PM causes permanent cost increases.
- o Production loss and maintenance costs should be optimized.
- o Breakdown and preventive maintenance costs should be optimized.

GUIDELINES FOR IMPLEMENTING A PM PROGRAM

- o Assign the most competent people to design PM system.
- o Implement in criticality order.
- o Implement incrementally to allow time to reap benefits.
- o Optimize PM versus BM per Chapter 13.
- o Optimize production loss costs versus maintenance costs per Chapter 4.

The following paragraphs give examples of the different choices available when making the decision on the degree of thoroughness varying from zero preventive maintenance to a preventive maintenance program that is designed for zero failures.

100% BREAKDOWN MAINTENANCE

This equipment is handled by the trouble call system. Nothing is done unless someone calls the trouble desk to announce equipment failure. An example might be an exhaust fan in a conference room with lifetime lubricated bearings.

BREAKDOWN MAINTENANCE WITH MINOR LUBRICATION ONLY

This case is the same as the preceding paragraph, except that the fan bearings require periodic lubrication. The lubrication would be handled as part of a preventive maintenance program. A car owner who plans to buy a new car every year could use this system performing no preventive maintenance other than checking the engine oil and adding oil when necessary.

MINOR PREVENTIVE MAINTENANCE

This type of maintenance consists of:

- Necessary lubrication
- Preventive maintenance inspection, adjustments and repair limited to a maximum time per machine of say, ten minutes. Anything over this time will be reported to the foreman and be covered by a special work order

The number of trouble calls and reports of problems during the inspection serve as an indicator of the adequacy of preventive maintenance instruction.

Many maintenance experts say that PM should be restricted to only this type of work. I say that advice is nonsense and that the type is just one of the choices you have and that you must select the type most appropriate to the situation.

MAJOR PREVENTIVE MAINTENANCE

This system is normally used on critical equipment and includes in addition to the items listed in Minor Preventive Maintenance, periodic parts replacements and periodic overhauls.

PREVENTIVE MAINTENANCE DESIGNED FOR ZERO FAILURES

This system uses in addition to the items listed in major preventive maintenance, preventive maintenance instruction optimization, Analysis of Critical Problems (see Chapter 11) and the Principle of Simultaneous Downtime (see Chapter 12). Two examples of equipment falling under this category from the author's personal experience, are all facilities designated launch critical at Kennedy Space Center and two 36' diameter x 300' long rotary kilns in a cement plant at a time when the plant was operating 24 hours per day, 7 day per week and when all of the plant's output could be sold. (Downtime would therefore result in an irretrievable loss in sales dollars).

A case study which shows an example of equipment needing zero failure PM follows:

CASE STUDY 1H: The Launch Critical Diesel Generator Set

At one plant, we had to retrofit 6 diesel - generator sets to provide power with tight specifications on voltage and frequency fluctuations and an allowable outage time of less than one second. The utility company power could not meet these specifications. The power was for critical computers and deviation from the specifications would cause a long list of catastrophic occurrences.

For example, if the power failed when the Apollo Spacecraft was on the launch pad during a launch countdown, several catastrophes might occur, i.e.:

- o Computer tapes might break at high speed.
- o If fuel was being loaded everyone would be operating in the blind and the countdown would have to be aborted and reversed under emergency conditions.
- o If the astronauts were in the missile, the countdown would have to be aborted and the astronauts evacuated under crash emergency procedures because nobody would know what's going on and the life of the astronauts would be in jeopardy.

As was the case with all critical equipment, we designed the preventive maintenance and operating systems for zero failure. Management, as usual, was willing to forgive one failure, but the second one was not permitted with the threat of being fired if the second one occurred. On one particular night, the operator had just shut down one set because it was making threatening noises; started up a spare; and was going through the synchronizing procedure. When he closed the circuit breaker, it blew up in his face and kicked all generators off the line. The previously mentioned catastrophes occurred and we worked furiously to get the set back on the line to minimize the delay. The problem was that we were forced to start up the generators without knowing what caused the explosion in the first place. After frantically investigating the mishap for one week, we were no closer to the solution and were afraid that the same thing would happen again. Although the operator had sworn that he had properly followed the paralleling procedures, it was concluded that the only possible answer left was operator error. When the operator was taken aside privately and offered immunity from disciplinary action, he confessed and the problem was solved. This incident taught two valuable lessons. These are:

- o When a serious mishap occurs, there is no excuse for allowing a recurrence. An investigation should be started immediately with the goal of finding the cause and preventing a recurrence. (See Chapter 11, Analysis of Critical Problems).
- o A problem knows no organizational lines. The problem could be in any department, not only maintenance, and may not be exclusively a technical problem. It could, for example, be administrative, communication, personnel or a morale problem. In this case, the problem was with the operating department and was a personnel problem.

The operator had been thoroughly trained in the synchronizing procedures. Furthermore, the instructions were printed in very large clear letters directly above the paralleling switchboard. The thing that we did a poor job of was explaining in detail to this man the serious consequences of a failure such as this one, with the result that he was careless in carrying out the procedure. All he had to do was to take his time and follow the printed instructions located right in front of him on the paralleling board.

PREDICTIVE MAINTENANCE (OPERATIONAL MONITORING)

This type of maintenance involves measuring selected operating parameters of the machine and shutting the machine down for repair only if the measured parameter goes outside of tolerance. Two examples will illustrate this technique.

1 - Aircraft Maintenance

Commercial aircraft engines formerly were maintained on an operating hour basis. You would do certain things after 50 hours, 100 hours, etc. and then completely overhaul the engine say after 5,000 hours (this is still the method with light planes).

Then, as I understand the history of engine maintenance for jet engines, they found that engine wear was more proportional to the number of take-offs, since at that time the throttle was opened full and the engine was under maximum stress. So they changed to that system and saved large amounts of money in engine overhaul as a result. Finally, they discovered that at times they were overhauling engines that seemed to have nothing wrong with them so they changed to operational monitoring (predictive maintenance). This can best be described by imagining the following allowable parameters on the fuel pump:

- o 100 GPM plus or minus 10%
- o 50 PSI plus or minus 5%
- o Vibration: XXX plus or minus 10%

Probes would be permanently installed on the fuel pump and the data would be transmitted to a black box which would schedule the plane for an outage and fuel pump replacement at its next stop, say at Los Angeles if any of the parameter tolerances were exceeded. This system saved them even more money since it allowed the pump to operate even longer. This principle uses the old law of maintenance attributed to Confucius who said:

"If it works don't mess with it."

2 - Plant Predictive Maintenance

This method is really the same as the aircraft maintenance example. For example, on a plant boiler it is quite common to have probes measuring boiler pressure, temperature, flow, etc. What is new is the ever increasing practice of extending this concept to more and more of the critical equipment in the plant by making rounds on the machines and taking readings with sophisticated measuring instruments such as vibration analyzers, infra-red detectors, etc., etc. These readings are frequently picked up by portable computers and then downloaded into the main maintenance computer for observation of trends. Then you can actually see the parameter approaching the limit, thus giving you advance warning of when you will have to shut the machine down for repairs.

In summary the types of preventive maintenance systems that are available for your use, as best fits the need of the particular situation, are shown in the following table:

TYPES OF PREVENTIVE MAINTENANCE

- 100% breakdown maintenance (BM)
- BM with lubrication only
- Minor PM
- Major PM
- PM for zero failure
- Predictive maintenance (operational monitoring)
- Combinations of any of the above

The astute maintenance manager will carefully select the type that best fits the application.

Part 2

The Downtime Improvement Cycle

8

General

Several things must be understood about the relationship between downtime and production lost costs, i.e.:

- o Downtime does not always cause a production loss.

Reasons for this fact are:

- o Redundant equipment is available, such as two electrically interlocked pumps so that if one fails, the other starts up automatically.
- o Equipment is producing into in-process storage so that other production. processes can draw from the stockpile while the equipment that broke down is being repaired.
- o There are product stockpiles in the external distribution system so that sales orders can be filled from the stockpile.
- o Plant is not operating 24 hours per day, 7 days per week, therefore, loss can be made up by working overtime.
- o Downtime on production equipment may be caused by items other than equipment breakdown, i.e.:

- Lack of tooling
- Lack of operators
- Lack of raw materials
- Operator idle time
- Lack of orders
- Etc.

o Downtime on equipment chargeable to maintenance may be due to several causes (see Figure 2-1).

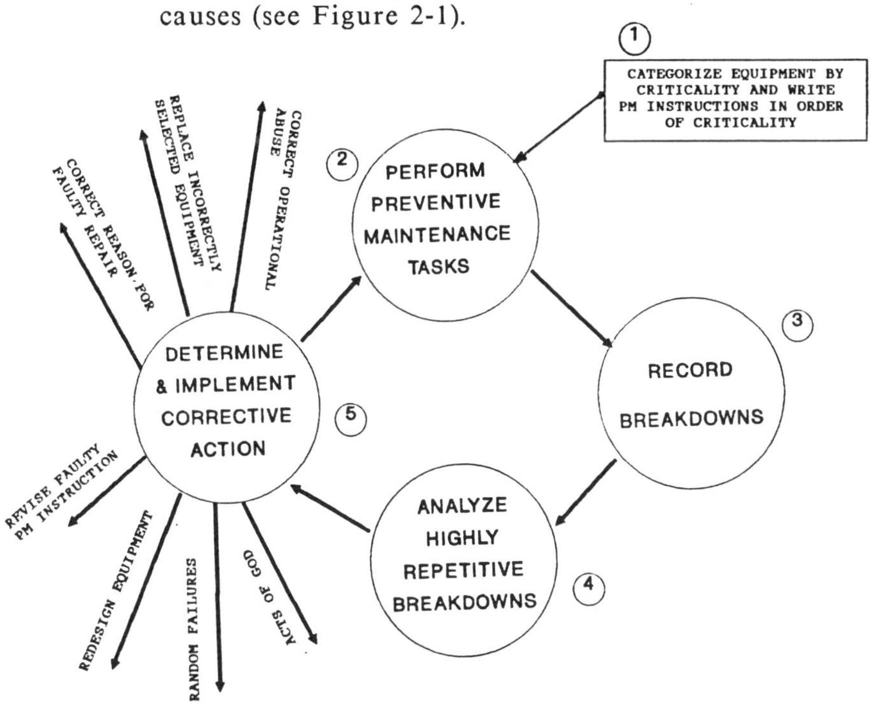

Figure 2-1: Downtime improvement cycle.

o Downtime should be recorded and charged to the department at fault, i.e., Maintenance, Production, Tooling, Material, etc. (See Case Study 1C on page 15.)

o Even if downtime does not result in a production loss cost, it usually causes a loss in productivity because people may be idled by the breakdown.

Finally, it can be said that the degree of loss from equipment downtime can be listed in order of severity:

1. An irretrievable loss in sales.
2. A production loss is suffered but it can be recovered by some means, for example, by the use of overtime.
3. A productivity loss is suffered by some department(s).
4. A maintenance productivity loss is suffered as a result of not using the downtime improvement cycle.

Downtime control can be achieved through a repeating Downtime Improvement Cycle (DIC). This cycle is shown graphically in Figure 2-1. Downtime is reduced each time the cycle is repeated. This cycle consists of the following steps:

- o Categorize equipment by criticality and write preventive maintenance instructions appropriate to the criticality for selected equipment.
- o Perform preventive maintenance.
- o Record equipment breakdowns. Breakdowns should be sorted by equipment part that failed, reason for failure and corrective action taken to correct the failure.
- o Analyze highly repetitive failures, determine cause and take corrective action to prevent recurrence.
- o Implement corrective action. Such action usually falls in one or more of the following categories: redesign equipment; correct operational abuse; replace incorrectly selected equipment; correct reason for previous faulty repair; correct random failures (this includes unexplained chance failures, for example, a foundation bolt on a machine. The

bolt is not supposed to fail, but it does. Another example is a computer chip. It usually fails, if it is to fail, on the first startup. If it passes this first startup it usually lasts for a very long time. There is nothing much you can do about these type failures except to use the zero failure PM system previously described); and, finally, revise preventive maintenance instruction. The last item closes the loop in the cycle.

Most companies follow a two-step procedure in handling equipment failures, i.e.:

- o Determine what failed.
- o Fix it.

The competent maintenance manager must in addition follow the DIC and determine the cause and prescribe corrective action that will insure that the unplanned failure does not repeat itself.

IT IS IMPORTANT TO RECOGNIZE THAT USE OF THE DIC WILL NORMALLY ELIMINATE THE PROBLEM PERMANENTLY. FAILURE TO USE THESE CONCEPTS CONSTITUTES INEXCUSABLE MALPRACTICE ON THE PART OF THE MAINTENANCE MANAGER.

The pages that follow will discuss the various steps in the downtime improvement cycle, starting with the recording of breakdowns. The reason for starting at this step is that this is the first system one would design and implement in an existing plant having no part of the downtime improvement cycle implemented.

9

Repair Records

GENERAL

The work order form is *the* document used to record equipment maintenance tasks and provides *the* most important source of information to facilitate maintenance cost and failure analysis and the writing of PM instructions. In order to show how the work order is used as a maintenance record, a specific work order form that was designed is described, fully recognizing that each company will have its own particular needs which may dictate a different type of form. Particular emphasis is given to the work performed box, which is where the repair record goes, and the failure code, which is the device for entering this information into the computer.

WORK ORDER PURPOSE

The basis purposes of the Work Order System are as follows:

- o To provide a means for screening and authorizing work.
- o To provide cost data segregated in a logical manner.
- o To provide feedback information on repetitive failures for analysis purposes.

- o To provide a tool to facilitate planning and scheduling of maintenance work.
- o To facilitate control of productivity.

WORK ORDER TYPES

A typical work order form is shown in Figure 2-2. This form is used in two ways, as a *Blanket Work Order* and as a *Special Work Order*. The two ways that the work order is used are described below.

******** MAINTENANCE WORK ORDER ********			
WORK ORDER #	MACHINE #	MACHINE DESCRIPTION:	
JOB TITLE:			PRIORITY #
REQUESTING CC:	PERFORMING CC:	REQUESTORS NAME:	REQUEST DATE:
WORK REQUESTED:			
CREW SIZE		TRADE	TRADE HOURS
EST	ACT	REPAIRMAN WELDERS ELECTRICIANS OTHERS	EST ACT
EST LABOR $	EST MAT'L $	EST TOTAL COST $	
ACT LABOR $	ACT MAT'L $	ACT TOTAL COST $	
APPROVED TO PROCEDE BY:			
WORK PERFORMED:			
FAILURE CODE:	FAILED COMPONENT NAME:	WORK TYPE CODE:	
COMPLETION APPROVED BY:		COMPLETION DATE:	

Figure 2-2: Work order form.

Blanket Work Orders

Blanket work orders are used for two purposes.

1. On routine repetitive small jobs, when the cost of processing an individual work order may exceed the cost of the job. For example, a trouble call calling for the replacement of a single fluorescent light tube. A blanket order to cover these tasks might read: "To cover the cost of replacing light bulbs in the administration building."

The blanket work orders are normally written only once per year. When a man does work under this order, he merely charges the work order for the time he spent at the end of the day, thereby saving the time and cost of filling out and processing a special work order.

2. When a job is a fixed routine, such as a janitor's job. In this case, the janitor does the same thing every day. His job is repetitive and pre-planned. You know what the janitor does beforehand and the reporting of what he does on individual work orders by task would result in useless and costly information.

In both cases, the blanket work order provides a gross cost for the generic task in cumulative to-date and annual costs.

Special Work Orders

A special work order is written for all other individual jobs. Implied in the writing of a special work order is the fact that the particular individual job is important enough to warrant separate individual approval and reporting of all pertinent facts about the job.

STEPS IN FILLING OUT THE WORK ORDER

The work order is filled out in three steps. These are:

Opening the Work Order

Certain preliminary information is put in the work order when it is first opened (the request stage). This includes the following information:

- o Work order number
- o Machine number
- o Machine description
- o Work requested
- o Requesting cost center
- o Performing cost center
- o Requestor's name
- o Request date
- o Priority
- o Job title

Planning and Estimating Information

The work order is then sent to the planning and estimating group who adds the following information:

- o Estimated crew size and hours
- o Estimated material cost
- o Estimated labor cost
- o Estimated total cost
- o Approved to proceed by:

In certain cases the planning and estimating stage may include engineering and drawings in which case the following is added to the work order as a supplementary package:

- o Drawings
- o Bill of materials
- o Lists of tools and equipment
- o Outage approval
- o Electric circuit assignment
- o Etc.

Work Order Completion Stage

When the work order is complete, the following information is added:

- o Work performed
- o Job title (update)
- o Failure code
- o Failed component name
- o Actual material cost
- o Actual labor crew size and hours
- o Actual labor cost
- o Actual total cost
- o Completion approved by:
- o Completion date
- o Work type code

HOW TO FILL OUT THE WORK ORDER FORM

Work Requested

Insert a description of work requested. The description should be as explanatory as possible. For example, if a production foreman noticed something wrong with a pump, this box might read "Please fix bad leak in pump shaft seal." If the job was a pre-planned job written by the chief-planner scheduler, the box might read "Perform complete overhaul on pump. Renew seals, all gaskets, broken nuts and bolts, clean thoroughly inside and outside. Renew gear; if necessary, repaint."

Machine Number

mber. It is suggested that 12 digits be
ine number. This number should be
ant's needs, however, a recommended
number follows:

Criticality code (see pages 25-27 of the textbook)

Digits 2 & 3 - Equipment category

Digits 4 & 5 - Equipment subcategory

Digits 6 - 10 - Sequence no. (0-99999)

Digits 11 - 12 - Reserved for PM instructions (see Appendix)

The criticality code is necessary to enable sorting by this code, a very important convenience.

An example of the equipment categorization is:

Equipment category: AIR CONDITIONING SYSTEM

Equipment subcategory: COMPRESSOR (A component of the air conditioning system)

The equipment categorization codes will enable sorting by equipment types, a worthwhile tool. However, if you are not interested in this convenience, you can simply use the 12 digit number as a sequence number.

Machine Description

Enter machine description, such as "Storage Hall Crane."

Cost Center

Enter appropriate cost center numbers of the department requesting and performing the work.

Requested By

Enter signature of person requesting work.

Request Date

Enter date of request.

Work Order Number

Enter work order number. There are 10 digits for your use in designing a WO number. The design should suit each plant's needs, however, a suggested design is shown below:

Digits 1 & 2 - Month

Digits 3 & 4 - Year

Digits 5 & 6 - Area code (divide plant into 0-99 areas)

Digits 7-10 - Sequence number (0-9999)

Priority Code

The priority number is of major importance. It indicates about how much time is left before a repair need becomes a serious problem. A repair need becomes a serious problem when it is likely to:

- stop production;
- injure someone; and
- damage equipment.

If any of the three conditions are likely to occur momentarily, you have an *emergency* condition (Priority Number 1). If you must interrupt your weekly schedule, but not your daily schedule, to avoid the possibility of any of the three above conditions, you have an *urgent* condition (Priority Number 2). If you are able to schedule the work at least one week in advance, you have a *normal* condition (Priority Number 3). If you have one month's lead time, you have a *programmed* condition (Priority Number 4). *Fill-in* work has no current time requirement (Priority Number 5).

The priority system plays a vital role in communicating urgency in establishing criterion for manpower allocations and workload balancing. The chief-planner scheduler shall enter the appropriate priority code in the box provided on the work order.

It is useful in the case of special types of work to use the following alternate priority system:

classifications for the following

sign

- o Failure analysis
- o Preventive maintenance instruction writing

Priority Numbers

No.	Name	Criterion	Action
1	Emergency	Production will be stopped unless repaired immediately. An extremely hazardous condition exists. Equipment will be damaged unless repaired immediately.	Repair immediately. Paper work to follow.
2	Urgent	A serious safety hazard exists and must be repaired before the end of the week and on the next shift, if possible. Production will be stopped unless repaired before next week.	Interrupt weekly schedule and place on next available daily schedule.
3	Normal	A defective condition has been identified. This condition will most likely not stop production, cause damage, or injure someone if corrected during the next week to four weeks.	Interrupt monthly schedule and place on next available weekly schedule.
4	Programmed	Predetermined repetitive repairs, period inspections, major maintenance repairs and construction will normally have this priority.	Place on next available monthly schedule.
5	Fill-in	Work assigned this priority has little or no time requirement.	Complete as time permits.

Estimate

The chief-planner scheduler (CPS) shall enter a rough estimate of the labor required to do the job in the space provided.

It is not intended that an exhaustive and very accurate estimate be attempted, but that a reasonable estimate be made based upon sound judgment. This will naturally require that the chief-planner scheduler be a person of considerable experience in all types of work.

Great care shall be made in selecting the most efficient crew size for the job because of the great effect this has on labor efficiency.

A detailed discussion on estimating methods and the reasons for and use of estimates is made in Part 3, "Productivity Control."

As can be seen from the estimating box on the work order, two quantities are required for each trade used, crew size and hours required per crew, i.e.:

2 R 4

This means 2 repairmen are needed for 4 hours.

After the job is planned and estimated, the person authorized to approve the start of the W.O. signs the W.O., indicating approval to proceed. At this point he has all of the information he needs to approve (or disapprove) the job.

After the job is complete, the actual costs and hours are entered so that the estimate and actuals can be compared. If there is a big discrepancy, an investigation and corrective action should be taken.

Job Title

The chief-planner scheduler shall enter a short accurate description of the job for input into the computer. The title should be updated after the job is complete.

The reason for this step is to provide a meaningful job title in the computer reports.

For example, it is possible that the work requested box might read "Please fix slurry pump, it doesn't work." This is not a meaningful title. A meaningful title in this case might be "Replace motor overload elements," since it describes the work actually performed.

Work Performed

The foreman, upon completion of the job shall enter the actual work performed, including:

- o Part that failed.
- o What corrective steps were taken?
- o What caused the failure?
- o What preventive steps should be or were taken?

Because the failure of one part frequently causes a chain reaction of failures, the part failing first must be identified. The information put in this box will vary depending on the circumstances, i.e.:

Case 1 - Preplanned Work: When a job is preplanned in advance, for example, on a construction or modification job or on a planned overhaul, the work is completely described in the "Work Requested" box. The "Work Performed" box shall be left blank if the job is performed according to plan.

Case 2 - Preplanned Work - Changed: If in the course of performing a preplanned job the work is modified in process, then a description of the change shall be entered in the "Work Performed" box. For example:

Work Requested

Install a 50' - 0' saddle in #7 Belt.

Work Performed

Job cancelled due to production problems.

Repaired damaged splice No. XYZ.

Case 3 - Breakdown Job or Trouble Call: In the case of a breakdown job or a trouble call, the work requested is frequently, of necessity, vague, i.e.:

Work Requested

Fan #XYZ is vibrating badly, please repair.

After the fan is inspected and the fault isolated and repaired, the foreman knows the detailed cause and what was done. Since this represents valuable feedback information for use in maintenance analysis, it must be recorded. It should be recorded in the work performed box as follows:

Work Performed

South Main bearing on Fan #XYZ dry of oil and burned up.

Replaced bearing

The recording of accurate feedback information on failures is one of the *most important* functions of maintenance foreman. You can depend upon the fact that good feedback will *normally not* be put on the work order without a strong campaign to achieve this by the maintenance manager, including lots of instruction and follow-up. The following is a typical letter that I wrote for the maintenance manager's signature to all of his foremen to emphasize the importance of adequate feedback:

CASE STUDY 2A: Letter on Feedback

TO: ALL MAINTENANCE FOREMEN

FROM: John Doe, Maintenance Manager

SUBJECT: INADEQUATE FEEDBACK DATA ON WORK ORDERS

An audit of completed work orders has revealed that the quality of information contained on the work order specifying the cause of failure is inadequate.

THE OBTAINING OF ACCURATE FEEDBACK INFORMATION ON THE CAUSE OF MAINTENANCE FAILURES IS ONE OF THE MOST IMPORTANT STEPS IN IMPROVING MAINTENANCE PERFORMANCE.

The information that we need and insist on getting on all work orders written for repair is:

(1) A description of the problem.
(2) A description of the part that failed.
(3) The foreman's judgement as to what caused the failure.

Item (1) is to be written on the work order under the heading "work requested." Items (2) and (3) are to be written on the work order under the "work performed" box.

An example will illustrate what constitutes adequate feedback data.

Trouble: Pump XYZ won't start.

Cause: Fuse blown. Fuse size was 10 A, should be 30 A. Replaced with 30 A fuse.

The above tells what the problem was; what part failed, and why.

The following are examples of unsatisfactory feedback data:

<u>No Feedback on Work Order</u>

Pump XYZ won't start.
(no cause recorded)

<u>No Work Order Written</u>

Pump failed, work charged to blanket work order, no written feedback.

<u>Inadequate Feedback</u>

Trouble: Pump XYZ won't start.

Cause: Replaced motor. (This does not state what the problem was.)

Trouble: Forklift inoperative.

Work Done: Replaced brake linings, replaced tires, ignition coil, spark plugs, changed oil, put in new solenoid valve and wiring. (This work does not specify the cause of failure.)

In the future all maintenance foremen shall record proper feedback on work orders in accordance with the principles outlined in this letter.

If a maintenance task is performed which constitutes valuable feedback information, the work <u>shall not</u> be charged to a blanket work order, since this would result in a loss of information. In this case, a separate work order should be written.

Concerning the cause of failure, it is recognized that certain failures will require an exhaustive and lengthy analysis. It is not expected that foremen will perform such an analysis, but should state, "It is suggested that Engineering investigate this problem." On the majority of jobs, however, it is anticipated that foremen will not only know the cause, but how to prevent a recurrence. In this regard, it should be noted that failures can be caused by:

(1) Lack of preventive maintenance
(2) Operator abuse
(3) Incorrectly selected equipment
(4) Faulty part design
(5) Poorly performed repair work, etc.

If the foreman isn't sure of the cause, he may be able to give his best guess of the cause. This should also be put in writing on the work order.

Completion Approved

When the job is completed, the foreman shall sign his name in the completion approved box. The signature will verify that he has inspected the completed work and that the work has been done in accordance with the plan (or modified plan) and that it is of proper quality.

Completion Date

After signing completion, the foreman shall enter the date completed.

Failure Code

A four digit failure code shall be entered after the job is complete and after *all* other entries are made. It is important that a person knowledgeable about the entire maintenance program make this entry since the information will be the main tool for maintenance failure analysis and a valuable tool for writing and updating PM instructions. A description of this coding system follows.

The system starts with a 4 digit failure code. The failure code is designed as follows:

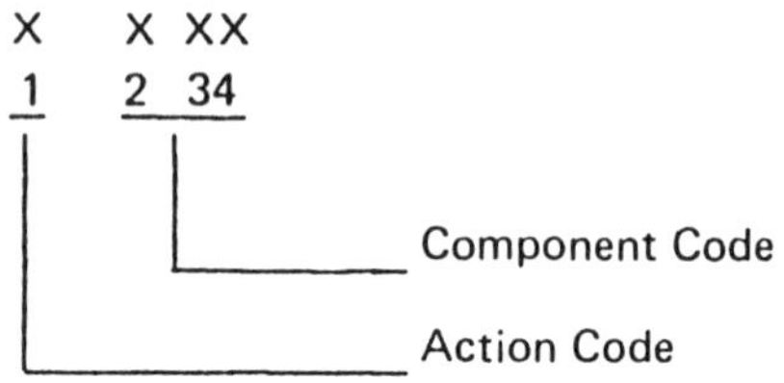

The definition of the Action Code is shown below. The definitions are self explanatory.

Code	Description
R	Replace
M	Repair
A	Adjust
C	Modification

These codes are a suggestion. Each maintenance organization may have need for more or different codes.

The last three digits are used to describe the component that failed. The following is a description of how to go about developing and updating this code:

Step 1: The code developer should call a meeting of all maintenance foremen and ask them to make a list of common component names used on equipment in their respective areas. For example, on a pump, the list might be:

Shaft	Nut
Impeller	Bolt

Packing	Housing
Impeller	Gaskets

This list alone includes components found frequently on many other equipment items.

Step 2: He then consolidates the list into one list eliminating duplications. This should include giving only one code to several component names meaning approximately the same thing (with suitable cross references). The more code numbers there are, the more difficult it will be to look one up and the more complex the failure report will be.

Step 3: Alphabetize the list.

Step 4: Number each component in sequence starting with 000 and going up to 999. This list will then be used by the work order coder for determining the last 3 digits of the failure code.

There will come the time when the person filling in the failure code will note that the component mentioned on the work order is not on the component list. He should then assign the next sequence number on the list to this new component.

This addition will result in the list not being 100% in alphabetical order and will therefore, make it increasingly difficult to find the item on the code sheet. This can be resolved by periodic submittal of the list to the computer department for alphabetical sequencing. This will result in there being two lists.

List 1: Sequential numbers, plus random component descriptions. This list will be used only for assigning new numbers.

List 2: Alphabetical component description list, plus random code numbers. This list will be used by the failure coder for finding the code number once the component name is known.

The purpose of putting the failure code on work orders will be described in the section on surfacing highly repetitive, critical failures in this chapter.

Work Type Code

A one digit code shall be put in this box to indicate the maintenance work type, i.e.:

Code No.	Maintenance Work Type
	Maintenance
1	Major
2	Preventive
3	Routine
	Breakdown
4	Emergency
5	Planned
6	Trouble call
	Changes to facilities
7	Construction
8	Modification
9	Rearrangement
	Services
A	Janitorial
B	Waste handling
C	Landscaping
D	Other
	Maintenance Engineering
E	Planning and estimating
F	PM instruction writing
G	Maintenance analysis
H	Scheduling
J	Production Loss (see page 65)

It is very important to know at all times the percentage each of the above categories is of the total labor plus material costs, since there is a reasonable percentage for each company. For example, the ratio of breakdown to planned maintenance, is a measure of the success or failure of the preventive maintenance program.

CASE STUDY 2B: Percentage Construction Work?

A request was made to do a study of the maintenance department in a copper mine and plant which seemed to be suffering from a steady decline in performance over the past 3 years. Since there was no recording of maintenance work type on work orders, the coding system described above was immediately implemented. A check of the percentages based upon the first month's trial run revealed that 75% of the workers' time was being spent on new construction.

A check of the company's history revealed that three years previous to the author's assignment, the company had become disenchanted with the performance of the firm that designed and constructed the plant. For two years they had been pushing the construction firm to clear up a long list of uncompleted and improperly designed facilities. In disgust, they finally threw the construction firm off the site and decided to complete the work on their own with the maintenance crew which they considered more competent.

They had never dreamed that this task would tie up 75% of their maintenance crew.

If they had coded the work orders described above, this fact would have surfaced immediately in the monthly computer runs, and they would have realized that they had "bitten off their nose to spite their face," and had made a disastrous error that had left inadequate numbers of maintenance workers to keep the plant running.

SURFACING HIGHLY REPETITIVE, CRITICAL FAILURES

Now that we have all of the information we need to analyze failures on the work orders, all we have to do is extract the information from the work order in such a way as to pin-point the most critical and most highly repetitive failures. A computer program has been designed to do this very simply. The program makes use of the following information that is on the work order:

- o Failure code
- o Equipment number and name
- o Criticality code
- o Work order cost
- o Date

With this information, the computer then prints the report shown in Figure 2-3. The report has the following features.

December 1975	Historical Breakdown Record	580101 S. H. Crane		
Component	Failure Record (Type of Maintenance, Date, Cost)	Number of Occurrences		Year-to-Date Cost, Dollars
		This Month	Year-to-Date	
013 Holding Cables	M01/01-200, M01/03-210, M01/19-199, R02/03-1000, R03/21-900, R04/01-1100, R05/06-1000, R06/16-1090, R07/12-1500, R07/12-1500, M08/19-80, M08/21-99, M09/23-102, R09/16-990, R10/18-1000, R10/11-800, R11/03-900, R11/04-600, R11/15-500, M11/21-60, M11/26-94, M11/28-92, M12/01-50, M12/15-100	2	23	12,666
069 Brakes	R01/16-400, R10/11-350, M11/15-40	0	3	790
169 Bucket	M03/17-50, M10/21-75	0	2	125

Figure 2-3: Failure analysis report.

- Report can be programmed to print out only those equipment items that are the most critical (or any other degree of criticality).
- Equipment printouts are made in descending order according to the equipment item having the highest number of failures.
- Failures are printed out in descending order of number of occurrences.
- Number of occurrences this month and year to date total are shown.
- Year to date cost is shown.
- The failure record shows the action code, the component, the transaction date and cost, i.e.:

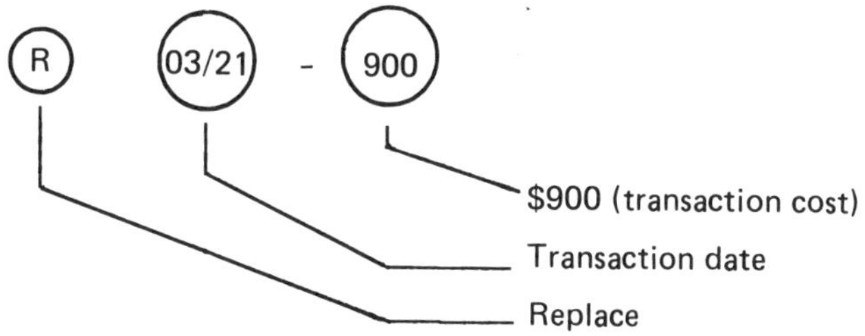

Thus, one can deduce from Figure 2-3, which is the first item on the printout that:

- Equipment has highest number of failures in plant.
- Holding cable maintenance and replacement is the biggest problem with the S. H. Crane, by virtue of the high number of failures and cost to date.
- There were thirteen replacements of the holding cables in one year, an astonishingly high and unusual failure rate.

This high degree of visibility saves the maintenance analyst an enormous amount of work. Most of the time, the mere surfacing of a highly repetitive failure leads immediately to its remedy.

As stated previously, it is a good idea to simultaneously look at the tab run on parts usage listed in descending order of annual cost as a cross check.

Only the December breakdown record reports should be filed (this is the annual report in effect). They should be filed in numerical sequence by equipment numbers.

Each month, maintenance will receive a monthly, year-to-date report consisting of historical records arranged in numerical sequence by equipment number. At this time, the previous month's report can be thrown away.

The record system will therefore consist of a monthly tab run discarded each month and a permanent file of annual reports for each equipment item.

Some plants contain many thousands of equipment items. This, coupled with hundreds of parts per equipment item, produces millions of components and a complexity factor capable of taxing the most competent plant engineer. He needs a tool to sort out the critical occurrence from trivia. The report shown in Figure 2-3 is such a device.

Assuming that a criticality code has been assigned to each equipment item, as described in Chapter 5, the plant engineer has the option to order the computer to print out criticality code A (the highest criticality) equipment only. This could mean that only 100 out of 20,000 equipment items are reported on.

The report itself identifies which failure is significant by virtue of numbers of occurrences, total dollars, interval between failures and type of failure.

The maintenance analyst, after determining which failure is significant, refers to the specific work orders to obtain more details on the failure in an effort to eliminate or reduce the numbers of failures. Progress in correcting failures can easily be determined by examining the trends on the December (annual) reports.

When the maintenance analyst feels that he has solved all the problems in criticality code A equipment, he can program the computer to print out "B" equipment and "C" equipment thereafter depending upon the size and capacity of the maintenance engineering staff assigned to maintenance analysis.

The beauty of the Figure 2-3 report is that as problems are solved, the next worst machine will automatically float to the top of the list.

In a large organization they frequently have area maintenance engineers. In this case it would be logical to assign these engineers to do the analysis on equipment in their areas since they will be more familiar with the problems in their area.

PRODUCTION LOSS AND DOWNTIME COSTS

In Chapter 6, the percentage that maintenance is of the total operating cost is discussed. If this percentage is high, the maintenance manager would be well advised to keep track of both downtime and production loss costs. As previously discussed, downtime can cost money by idling production workers, over and above the maintenance cost. Production loss costs may be suffered if a deadline machine causes a loss of sales.

Equipment, such as rotary kilns in a cement plant or a paper machine in a pulp and paper mill or a boiler in a utility plant, are so expensive to buy, and can cause such large losses when they break down, that a very complete record should be kept of all downtime occurrences, and the downtime or production loss costs should be charged to the department responsible for the downtime. Typical examples of the departments charged and causes are:

- Tool department--Lack of tooling
- Production department--Lack of operators
- Material department--Lack of material
- Maintenance department--Equipment failure

There are several ways to handle the accounting of these costs, i.e.:

- o Have the production operator keep the records right at the machine as a separate accounting system. The production department will then hold periodic meetings which show the percentage downtime by department and discuss ways to reduce the downtime.

- o Have the department that suffers costs as a result of equipment downtime, charge the maintenance department for the cost. This can be done by their writing a maintenance work order and using code J shown on page 60 for production loss costs. This system will isolate production loss costs and will put the pressure to reduce downtime where it belongs, on the maintenance manager.

If the machine is particularly critical, the maintenance manager may want to record downtime hours as well as costs. Foremen in this case would record the total amount of time that elapsed from the time a piece of equipment became inoperative because of an equipment failure to the time it was placed back into operation. This can be done after the job is complete in a special box labeled "downtime hours" on the work order.

This information will be used for two purposes.

- o To plot curves of downtime trends for use by top management as a control tool.

- o To provide additional information for foremen when analyzing repetitive failures.

When a foreman is analyzing a repetitive failure on an equipment item, he should be aware of the amount of time the equipment was deadlined. Obviously, it is a more serious matter if the equipment was out of service for four weeks as opposed to ten minutes. The equipment suffering four weeks' downtime should generate in the foreman a greater sense of urgency in eliminating the repetitive failures. In the case of the four weeks' downtime, not only should the foreman try to correct the repetitive problem, but he should look into the reason for the four weeks' downtime. The reason may be material procurement time in which case he should work on shortening this time. Perhaps some of the repair parts were not stocked, or they may be stocked and the MIN-MAX is incorrect resulting in frequent stock outages. It is also possible that the repair time is excessive, in which case he should consider repair methods improvement.

The following definition of terms will help in understanding the thorough analysis which is sometimes necessary to make on critical equipment:

> *Total Downtime (TD)*--Is the total time the equipment is out of service because it has broken down and is inoperable.
>
> *Production Lost Time (PLT)*--Is that portion of the TD that production was lost due to equipment failure.
>
> *Availability Time (AT)*--Is the normal number of hours per week that the equipment is planned to be available for use.
>
> *Deadlined (DL)*--Deadlined means that the equipment is broken down and unavailable for use.
>
> *Utilization Percentage (UP)*--Equals AT minus PLT divided by AT.

The goal naturally, is to make UP as high as possible. TD is of value in analyzing the total time required for repairs. Naturally maximum use should be made of the time when the machine is not required for production to effect repairs. If the machine takes only one hour to repair and the TD is 80 hours, the analyst should determine the reason for this anomaly.

The recording of production loss and downtime costs is necessary to optimize these costs with maintenance cost as described in Chapter 4. In addition, it will be shown in Chapter 13 that breakdown and preventive maintenance costs should be optimized regardless of whether or not there are downtime or production loss costs.

USE OF SUMMARY REPORTS

Summary reports can be of great use to the maintenance manager. The information for these reports is obtained from the work order system and is printed in useful form by the computer. Examples of useful summary reports are:

- o Maintenance cost--$/Ton
- o Maintenance cost--$/Unit produced
- o Annual maintenance report shown in Figure 1-4
- o Janitorial costs in cents per square foot
- o Costs by work type code, including percentage each type is of total
- o Parts use report arranged as in Figure 1-4

The reports should be subdivided by maintenance organization units. Standards and improvement goals should be set on the above indicators. Figures outside of standard should be further investigated by referring to the individual work order and in the case of failures, by referring to Figure 2-3 and then to the work order. Other examples of such reports follow.

(a) Annual Equipment Report

This report summarizes work order costs for the year by equipment number and is given in the following format:

Equipment No. xxxxxx

Work Order No.	Description	. .Labor . . (hr)	($)	Parts ($)	Outside Service ($)	Total ($)
x-xxxx	Replace drive motor	xxxx	xxxx	xxxx	xxxx	xxxx
	Totals	xxxx	xxxx	xxxx	xxxx	xxxx

This report should be printed in descending order of the total dollars column.

(b) Monthly Year-To-Date Major Problem Reports

The following two reports are recommended as a quick and easy means of surfacing principal maintenance problem areas:

Report #1

1 Equip. No.	2 Description	3 YTD Downtime Occurrences	4 YTD Maintenance Costs	5 Percent Column 4 to Total Column 4
xxxxxx	- - - -	xxxx	xxxx	xxxx

Report #2

Equip. No.	Description	YTD Maintenance Costs	YTD Downtime Occurrences	Percent Column 4 to Total Column 4
xxxxxx	- - - -	xxxx	xxxx	xxxx

All line items should be listed in decreasing magnitude of column no. 4.

The preventive maintenance supervisor should be *particularly* on the lookout for equipment items that account for a *major* percentage of costs or downtime occurrences. For example, if the two reports showed that only ten equipment numbers accounted for 90% of the maintenance costs and downtime occurrences, then quite obviously these ten items should receive close to 90% of the maintenance analysis time. The intent is to prevent squandering of valuable analysis effort on an across-the-board effort with proper relationship between effort and return on effort.

It should also be emphasized that maintenance cost and downtime are *not* the only barometers of an effective maintenance program. There are many other factors that must be considered, for example, production loss and safety, to name a few.

(c) Monthly Parts Usage Report by Equipment

Another valuable tool for surfacing maintenance problems can be obtained by arranging the annual issues per stock item in descending order as follows:

. . . Equipment Number: XXXXXX . . .

ACC #	Description	No. of Work Orders	Total Quantity Issued	Average Unit Price	Total ($)	Percentage
XXX1	Gears	X	X	X	3,000	67
XXX2	Rollers	X	X	X	1,000	22
XXX3	Bearings	X	X	X	460	10
XXX4	Rope	X	X	X	20	0.6
XXX5	Nuts	X	X	X	10	0.2
XXX6	Bolts	X	X	X	10	0.2
	Total				4,500	100

This report clearly shows that the major parts problem with this machine is *gears.*

10

Failure Analysis

GENERAL

The first step in the failure analysis procedure is to select the most critical machine (using the criticality code) that suffers the biggest downtime problem (using Figure 2-3).

There are several types of failure analysis systems. Each type has its place. The types are shown in the following list:

TYPES OF FAILURE ANALYSIS

TYPE	*COMMENT*
1-Analysis during emergency repair	Should be small percentage
2-Analysis of repetitive failures	Bulk of analysis (Chapter 10)
3-Analysis of critical problems	For major catastrophes (Chapter 11)
4-Principle of simultaneous downtime	For large complex equipment (Chapter 12)
5-Analysis to achieve zero downtime	When loss of life or large production loss is threatened (Chapter 12)
6-Paper analysis	When solution is uncertain and risk is high (see case study 2C)
7-Annual shutdown and overhaul	For continuous operation and when idle period is available

This chapter will discuss principally types 1, 2 and 6.

After we select the failure analysis type, then we should gather all of the information available that is written on the failure, and later on to supplement this with personal interviews with persons having knowledge of the failure. Written information that can be used is:

- Summary reports
- Repair records (see Chapter 9)
- Work orders
- Drawings
- Preventive maintenance instructions
- Modifications
- Correspondence files
- Purchase orders
- Stock records
- Analysis of critical problems (see Chapter 11)
- Principle of simultaneous downtime studies (see Chapter 12)
- Manufacturer's data
    - Operating instructions
    - PM instructions
    - Trouble analyses charts
    - Spare parts lists

If an analysis of the above written data fails to solve the problem, then an interview should be conducted with one or all of the following persons who may have knowledge of the failure:

- Maintenance mechanics
- Machine operators
- Maintenance and production foreman
- Planner/schedulers
- Engineers

USE OF THE FAILURE ANALYSIS CHART

The Failure Analysis Chart (Figure 2-3) is one device for isolation of equipment problems. Several observations are made relative to the use of this chart.

- o The depth of the column for each component failure gives an immediate visual indication of which failure occurs more frequently.
- o The frequency of occurrence is summarized in the table by month and year to date.
- o Another item which shows up is the time between occurrences. This must be weighed with the other facts. Something occurring ten times per month is obviously more serious than ten times per year.
- o A further indication of the magnitude of the problem is given in the year to date cost column.

After it has been determined that a problem is worthy of analysis by reason of the equipment criticality, the high repetitiveness of the failure or the high cost, a more detailed study must be made.

The first step is to obtain from the file all of the work orders listed on the Failure Analysis Chart for the particular failure being studied.

If the feedback on the work order is adequate, the detailed explanations on the work order should further isolate the cause of the failure.

Furthermore, the foreman's opinion of what the cause was will be recorded. If the feedback is inadequate, the analyzer will want to talk to the foreman and any other person knowledgeable of the problem and he will want to examine files containing drawings, correspondence, purchase orders, modifications, etc. on the equipment.

In the case of multiple failures, the analyzer must isolate the original part that set up the chain reaction (pyramiding failures).

DETERMINING THE REASON FOR FAILURE

Once a determination has been made as to which detailed part failed, the next question is why did it fail? The answer will be found under one or more of the following general classifications (see Figure 2-4):

- o Lack of preventive maintenance
- o Improper frequency of preventive maintenance
- o Faulty previous repair
- o Incorrect equipment selection
- o Faulty design
- o Faulty manufacturer
- o Operator abuse or error

Item	Cause of Failure	Corrective Action
1	Faulty Design	Redesign
2	Faulty Manufacture	Warranty Action
3	Incorrect Equipment Selection	Modify Equipment Replace Equipment
4	Incorrect Operating Practices	Correct Operating Errors
5	No PM	Add PM
6	Incorrect PM	Correct PM
7	Faulty Breakdown Repair Work	Train Maintenance Craftsman

Figure 2-4: Cause of failures.

The biggest error that can be made at this state is to say, "Oh well! It's just one of those things, parts will fail!" Or some similar statement indicating a lack of interest in the reason for failure.

Learning the *reason for failure* in detail is the "name of the game" in maintenance improvement. Foremen and the maintenance analyst must be indoctrinated in being inquisitive about the reason for failure on all breakdowns they are involved in.

Maintenance analysis and optimization functions should be directed towards achieving one of the following levels of maintenance effort:

- Highest level--Zero downtime
- High level--Least possible downtime at the least cost
- Normal level--Most economical maintenance point
- Zero level--100% breakdown maintenance

The determination of which level to select is dependent upon factors discussed in Chapters 7 and 13.

It is a fact that the solution to the *majority* of repetitive failures is automatically found when the reason for failure is discovered.

For example, if a hydraulic or pneumatic hose system is failing much too frequently, the following questions might be asked.

- What was the design pressure?
- What was original hose and fitting specs?
- What was actual pressure?
- What was actual hose and fitting specs?
- Were incorrect hose and fittings used?
- Were they installed correctly?
- Did the hose fail from overpressure, old age, erosion or corrosion from inside or outside?
- Is the hose subjected to excessive vibration?
- Is the hose subjected to continuous movement?

These are some of the questions that might be asked. The answers to these questions should lead to the reason for failure.

DETERMINING CORRECTIVE ACTION

As was stated previously, the majority of solutions to highly repetitive problems are found automatically the same time the reason for failure is determined. To demonstrate this, a few actual examples are given.

PROBLEM 1.

Frequent hydraulic hose failure

Cause:

Hose subjected to overpressure due to faulty operation of pressure reducing valve.

Solution:

Repair reducing valve. Install relief valve on downstream side set to protect hose against overpressure.

PROBLEM 2.

Frequent pump outages.

Cause:

Fuses blowing in control circuit. Fuse was incorrect size and type.

Solution:

Installed correct size and type of fuse.

PROBLEM 3.

Frequent hydraulic hose failure on *all* vehicular heavy equipment.

Cause:

Foreman has substituted 2-ply plain hose with manually installed fittings instead of original design which was 5-ply armoured hose with machine swedged fittings because new hose was easier and simpler to install and the company did not have a swedging machine.

Solution:

Purchase swedging machine. Use specified hose and fittings.

The solution to some problems may be above the skill or responsibility level of the maintenance foreman either because of technical complexity, large dollar amounts required for the fix or because of the need for a process modification. These problems should be handled by the maintenance or engineering staff as directed by the maintenance manager. In some cases, the problem may not be solvable by the company's staff. If the consequences of the repetitive failure are serious enough by reason of cost and lost production, consideration may be given to the use of outside consultants.

In summary:

WHEN:

- o Solution is too complex or time consuming
- o Zero failure is mandatory
- o Process modification is required

THEN:

- o Refer solution to maintenance engineering or research lab

An example of this type of problem is described in Case Study 2H. The analysis could include:

- o Destructive testing and analysis of failed components.
- o Redesign for increased reliability.
- o Spares reassessment.
- o Beefing up PM.

There are several pitfalls that the analyst must guard against in his problem solving function.

- The tendency of many analysts to "drop the ball" if *their* solution is turned down. When an idea is turned down, the following questions must be asked:
  - Was it turned down for cost or lack of funds?
  - Was it turned down because the approver didn't think it would work?
  - Can it be done cheaper?
  - Can it be done differently?
  - Can the scope of the solution be reduced to save cost?
  - Was the savings justification adequate?
- The tendency of some analysts to "drop the ball" if the fault lies outside of their department, i.e., operational abuse, incorrect equipment, problem is electrical, or a training or personnel problem.

  Analysts must be prepared to cross organizational or functional lines when necessary to solve a problem.

The process of analyzing collected data can be summarized as follows:

- Consider reasons for failure (see Figure 2-1).
- Solution to problem frequently found just by reading material in files (read Case Study 2K).
- Corrective action usually obvious when reason for failure is determined.
- If corrective action isn't obvious, then all data must be analyzed.
- Determine corrective action.

EQUIPMENT OUTAGES AND RELATIONSHIP WITH PREVENTIVE MAINTENANCE

Remembering that our maintenance goal is to achieve the lowest combined cost of maintenance plus production loss, it is important for the maintenance analyst to carefully consider the question of equipment outages.

Outages facts that the foremen must know in order to minimize outage time during repair:

- Equipment causing production loss when inoperative (plan to minimize downtime hours).
- Equipment possessing redundancy (allows time for repair.)
- Equipment producing into stockpile and size of stockpile allows time for repair.
- Equipment downtime record (knowledge of this facilitates corrective action).
- Details of PM program on equipment (any weaknesses?).

The manner in which equipment outages are handled is important for several reasons:

- It affects production loss cost
- It affects crew efficiency
- It affects the ratio of production loss to major overhaul cost

The handling of equipment outages can be divided into two categories:

- Emergency Outages
- Preplanned Outages

Emergency Outages

The most important first step during an unanticipated breakdown is to arrive at a decision of what to do about the failure. Run the equipment as is? Make a patch repair?

Make a proper fix? Fix other things as long as the equipment is down?

The time required to make this decision causes production loss and a productivity loss for the maintenance crew. The decision of *how* to fix it also affects production loss. It is important that the decision be made *quickly* and *correctly*. This requires the collaboration of a knowledgeable person from maintenance and production. A perusal of hundreds of critical problem reports from other companies shows that errors in judgment concerning this subject are not uncommon.

The analyst should:

- Discuss the subject with maintenance and production foremen periodically to sharpen their sensitivity to the problem.
- Consider this subject in the writing of critical problem reports. Poor performance in this area surfaced by the critical problem report should be made the basis for corrective action including briefing of foremen on the problem and solution.

The failure analysis during an emergency repair should proceed in the following order:

- If the equipment is non-critical, the emphasis should be to find out what failed, fix machine and get it back into operation. Use trouble analysis charts showing trouble, probable cause and remedy provided by manufacturer.
- If foreman knows reason for failure, he should record this on the work order.
- Standard analysis will be done later. If machine is critical, do critical analysis immediately (Chapter 11).

Preplanned Outages

Preplanned outages for maintenance on production equipment running intermittently or with spares or in-process storage arrangements are no problem as far as production loss is concerned. It is the equipment that is running around the clock that is the problem.

There are several things which can and should be done to optimize the relationship between downtime for *preventive* maintenance purposes and production lost time cost. These are:

- Identify equipment which may cause a production loss when inoperative.
- Identify if equipment is producing parts into a stockpile and how long the stockpile will last. (This time may allow downtime without a production loss.)
- Record existing annual preventive maintenance program on the equipment and its statistical performance including:
    - Maintenance cost per ton or per dollar of product output
    - Downtime hours per year
    - Downtime percentage
- Determine if there is redundant equipment.

These four items represent what the present maintenance plan is and its performance. The next step is to outline several alternate plans and *estimate* their probable performance characteristics. Those plans which appear very good on paper may be worthy of an actual test. Alternative methods which may be considered are:

1. **100% Breakdown Maintenance**

2. **Light Preventive Maintenance Plan:** This includes light preventive maintenance work on items that can be worked on while running or appurtenances that can be shut down or that are provided with backup.

3. **Heavy Preventive Maintenance Plan:** This includes a major overhaul at some logical frequency which depends on the equipment, i.e., annual, semi-annual or bi-annual.

An example of this plan is the method of maintaining reciprocating aircraft engines. This plan includes certain preventive maintenance functions that must be performed after 100 hours, 500 hours and 3,000 hours of operation. The intention of this preventive maintenance program is to achieve zero downtime.

4. Operational Monitoring: The best example of this plan is the present method of maintaining jet engines on many of the top airlines. This plan includes the installing of instrumentation which measures the performance of the components under the plan. For example, the fuel pump might be instrumented to record fuel pressure, flow, bearing temperature and vibration. These measurements are continuously recorded and periodically transmitted to a central computer for analysis. If the operating characteristic exceeds the tolerances on the standard, an order is issued for a repair or replacement. In other words, no maintenance is ordered unless there is a deviation from operational standards and tolerances. It is reported that this plan has more than doubled the operating life of the engines and made a drastic reduction in maintenance cost.

5. Combination Plans: Finally, combination plans, using any or all of the above methods, may prove to be worthwhile.

Changes in the maintenance programs for very large, costly and critical equipment should be approached with caution because of the large costs involved. For this reason it is wise, if possible, to try and reason out the impact and results of the change on paper first. A well thought out paper analysis may not be exactly similar to actual operation, but may be close enough to see if it is an obvious improvement or not.

An actual example of a paper analysis and results should help in illustrating the point:

CASE STUDY 2C: Ball Mill Maintenance

Company

A mining firm.

Equipment

39-1000 HP synchronous motor drive ball mills.

Old Maintenance Plan

Variable frequency preventive maintenance.

Existing Historical Records

Firm had very complete historical records on failures.

Existing Average Downtime

15%

This company in analyzing its records noted that there was a random failure pattern averaging three months. An analysis on paper showed a huge savings in downtime if the random failures could be consolidated to one downtime occurrence.

The "mock trial" calculations were made as follows:

Existing Operating Hours

39 mills x 8,736 hours per year = 340,704 hours per year.

Proposed Downtime

Take 3 ball mills down for maintenance (plus appurtenances) for 8 hours once per week. Since there are 39 mills, it would take 39 divided by 3 equals 13 weeks to do this to all mills. Then the cycle would be repeated. This would result in each mill getting an 8 hour overhaul, 4 times per year.

During the mill outage all items would be open for inspection and faulty items fixed as far as is possible. If 8 hours was insufficient, the mills would be worked on around the clock until complete.

Downtime = 3 mills x 8 hr/day x 1 day/wk x 52 wk/yr =
1,248 mill hr/yr
Total mill operating hours =
39 mills x 24 hr/day x 7 days/wk x 52 wk/yr = 340,704
Downtime percentage = 1,248/340,704 x 100 = 0.4 of 1%

Results

The plan was put into effect and the downtime was reduced to 1% or only 0.6 of 1% over estimate. This was a reduction of 14% in downtime, a noteworthy improvement.

There was considerable resistance to this change since production didn't relish the idea of "unnecessary" planned downtime; unnecessary being interpreted to mean "Why shut down if it is still running?"

As the years went by with this new system, they found that certain items would always need replacement or repair during the shutdown. They would be ready with parts needed for these items. They would also experiment with new designs or materials on highly repetitive failures on one mill. If the new design or material was an improvement, the improvement was put on all mills.

When there was an uncertainty about a certain part, the question was asked "Will it last another 3 months? If the answer was "Yes," it was left alone and a note made to inspect it during the next overhaul. The results of this analysis was to reduce the planned downtime from 8 to 5 hours for each mill.

To summarize in the case of planned outages:

<u>PLANNED OUTAGES</u>

- o Planned outages should be designed for minimum outage time in all cases. Consider the use of the CPM method if the job is large and complex.
- o If outage for major repair or PM threatens a production loss, a special effort must be made to minimize downtime and, if possible, avoid a production loss, i.e.:
  - Use CPM.
  - Use kitting.

- Design special jigs, fixtures and material handling equipment.
- Provide detailed dismantling and repair instructions and drawings where they will help in cutting down overhaul time.

PRODUCT SPILLAGE

Product spillage, dust generation, litter and uncollected trash is another problem which may have unfavorable side effects not experienced in an ordinary equipment malfunction, i.e.:

Low Morale

Nobody likes to work in an unclean environment.

Production Loss

Product spillage may result in damage to equipment and eventually cause a shutdown.

Health Hazard

Uncollected fumes, dust from spillages, etc., may result in a hazard to health.

Safety Hazards

Product spillage may cause safety hazards.

Fire Hazards

Uncollected trash, papers, oily rags, or other combustibles may create a fire hazard.

Low Productivity

An unclean plant creates an atmosphere of disorderliness and confusion and may lead to low productivity.

Generates Further Uncleanliness

When a plant is not clean and neat, there is more of a tendency to make it worse. Nobody likes to throw a cigarette butt on a clean floor. However, if the floor is already a mess, nobody hesitates.

Product spillage and dust is just one of the manifestations of equipment malfunction and should, therefore, be handled in the same manner. This includes all phases of the downtime improvement cycle. Let us visualize a simple belt maintenance program, for example:

Preventive Maintenance

None

Breakdown

Product spillage through skirts

Analysis

Product is spilling through worn skirts and building up in a pile which submerges the belt and rollers. Estimated monthly breakdown cost is $500 per month and 16 hours downtime.

Corrective Action

Write preventive maintenance instruction to change skirts and scrapers once per month. Estimated cost: $150 per month.

In analyzing a spillage problem, consider the following:

- What component failed and what can we do to prevent reoccurrence of the failure and its consequences? (Let us assume that the failure is the worn and torn skirt; the consequences are product spillage, damage to the belt and rollers and equipment outage.)
- Why didn't production personnel notice the spillage and at least push it out of the way far enough to prevent the equipment damage and downtime?
- If the cause of the spillage is not going to be stopped immediately, what can be done to remove the spillage and prevent further equipment damage and downtime?

A spillage problem can be resolved either partially, or totally, by one or more of the above alternatives. The mistake that can be made in this respect is to try to solve the problem by means of one alternative and then to drop the entire investigation upon disapproval of this alternative without investigating the other alternatives.

IMPLEMENTING CORRECTIVE ACTION

It is anticipated that once the feedback systems are properly functioning together with the maintenance analysis function, a backlog of work will soon develop. The backlog of work should go through the following processing steps:

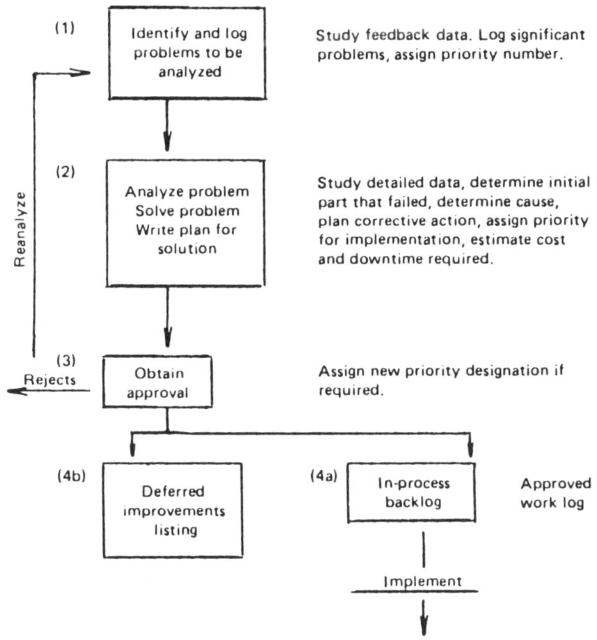

It is suggested that a log book be kept to record the magnitude of the analyzing workload and of the approved corrective plans. The desirability of the corrective effort can be prioritized by the assigning of a simple priority code as follows:

Code	Description
M	Solution is mandatory
HD	Solution is highly desirable
D	Solution is desirable

Solutions to problems which have a marginal ratio of benefits to cost can be either:

- o Rejected
- o Referred for further analysis
- o Logged in a deferred improvements log

An important part of implementation is the follow-up analysis to determine if anticipated benefits are achieved. Confirmation of benefits can be determined by watching the changes to the failure analysis chart each month. A complete elimination of the problem may be evidence of an "overkill." For example, a change from lubrication from once per year to once per month may eliminate a bearing problem, but may also cause unnecessary preventive maintenance costs. The correct period for any particular plant may have to be determined by trial and error.

MAINTENANCE ANALYSIS EXAMPLES

Examples of typical analyses are shown in Figures 2-5 to 2-9. The following comments are made relative to these examples:

- o Note that production loss cost is included as part of the cost of the maintenance failure. This is a *very* significant point. The reason it is significant is because this cost can at times dwarf the pure maintenance cost. As a result, it sometimes pays to spend *more* money on maintenance if it will reduce the production loss cost and, therefore, the total cost of maintenance.

For example, if the production loss cost is $50,000 due to a part failure and the chip replacement part costs 50 cents and the best and most reliable part costs $2, then buy the best part!

Maintenance Analysis Form

Historical Record: *Truck No. XYZ*
Clutch being replaced twice per year

Old Method:	New Method:
Operator was riding the clutch.	*Train operator in correct operating method.*

Item Description	Cost in Dollars		
	Old	New	Diff.
Labor: *Replace clutch $100 x 2*	*200*	*0*	*200*
(Assume car will be traded at 60,000 miles and that			
no clutch replacement will be required.)			
Material:			
Clutch $50 x 2	*100*	*0*	*100*
Production Loss Cost: *None*			
Total:	*300*	*0*	*300*
Downtime in Hours	*32*	*0*	*32*

Figure 2-5: Maintenance analysis example 1—incorrect operating practice.

o When comparing an improved method with the old technique, *it is usually not necessary* to make the comparison in writing as shown in the examples. Also, it is not necessary to be extremely accurate with the estimate. "Guestimates" will be sufficient. The purpose of the examples is to show the thinking process.

Maintenance Analysis Form

Historical Record: *Truck 2 PT*

Tires are replaced quarterly. Annual mileage is 24,000 miles.

Old Method:	**New Method:**
No Tire PM	*Worn tires are being caused by underinflation. Instructed operators to check tire pressures daily.*

Item Description	Cost in Dollars Old	New	Diff.
Labor: *Old*			
4 tires x 4 times/year = 16 tires x 10 min/tire			
= 160 min or 2.7 hrs			
2.7 hrs x $4/hr	*10.80*		
New			
10.80 / 4 = 2.70/yr		*2.70*	*8.10*
Material: *Old 16 tires @ $100.00 each*	*1,600*		
New 4 tires @ $100.00 each		*400*	*1,200*
Production Loss Cost: *None.*			
Total:	*1,610*	*403*	*1,202*
Downtime in Hours	*8*	*2*	*6*

Figure 2-6: Maintenance analysis example 2—no preventive maintenance.

- In cases where the old and new methods compare very closely, the comparison should be done in writing.

- All of the comparisons given in the examples are clear cut as far as whether the recommendation (New Method) is better. This is not always true, nor are the costs of a new method always known, nor are the consequences of a new method always predictable.

Maintenance Analysis Form

Historical Record: *Equipment No. XYZ*
Main fuses failing at rate of 10 per month.

Old Method:	**New Method:**
Use 10 amp fuses	*Investigation revealed design error. Fuses should be sized at 20 amps.*

Item Description	Cost in Dollars Old	New	Diff.
Labor: *Travel time to and from* 20 min.			
Diagnose trouble 5 min.			
Replace fuse 5 min.			
Total 30 min.			
@ $4.05/hr = $2.00			
10 occurrences x $2	20	0	20
Material: *Fuse – 10 amp, 10 x 50¢ = $5*	5	0	5
Production Loss Cost: *None – In process storage and redundant units available*			
Total:	25	0	25
Downtime in Hours	10	0	10

Figure 2-7: Maintenance analysis example 3–faulty design.

INDETERMINANT MAINTENANCE

As mentioned in the last paragraph, it is not always possible to predict exactly what will happen when a change is made in a maintenance task. For example, what will happen if you change the greasing of a motor from once per three months to once per four months?

Maintenance Analysis Form

Historical Record: *Floor Sweepers (1.0 of)*
Drive belts are failing on the average of once per 10 months

Old Method: *PM instructions on 10 floor sweepers calls for drive belt inspection once per month.*

New Method: *Belt inspection once per month uneconomical. Discontinue inspection and call for belt renewal semi-annually (since belts are now failing @ 10 months.)*

Item Description	Cost in Dollars		
	Old	New	Diff.
Labor: *Old*			
Inspection: 10 units x ½ hr per x $4/hr x 12 times per year	240		
Belt Replacement: 10 units x $4/hr x ½ hr/unit x 12/10	24		
New			
Belt Replacement – 2/yr x 10 units x ½ hr/unit x $4/hr		40	224
Material: *Old*			
Belts – 10 units x $5/set x 12/10 yr	60		
New			
10 units x $5/set x 2/yr		100	-40
Production Loss Cost:			
None			
Total:	324	140	184
Downtime in Hours	0	0	0

Figure 2-8: Maintenance analysis example 4–incorrect PM.

You think nothing different will happen, but you're not sure. In other words, the consequence of the change is *indeterminant.*

This is frequently the case when you are contemplating a change in frequencies. The only thing that can be done under this circumstance is to guess, and to determine action by this "guesstimate."

Maintenance Analysis Form

Historical Record: *Conveyor XYZ (Category A)*
Skirts and scrapers are being worn out and replaced 6 times per year.

Old Method: *Belt fasteners installed without countersinking of belt. The raised fastener is tearing skirts and scrapers.*	**New Method:** *Instruct craftsman to countersink all belt fasteners.*

Item Description	Cost in Dollars		
	Old	New	Diff.
Labor: *Old*			
6 occurrences			
4 hours/occurrence			
$4/hour	96		
New			
1 occurrence per year		16	80
Material: *Old* *6 occurrences*			
$40 for skirts and scrapers			
$20 for fasteners	360		
New *1 occurrence per year*		60	300
Production Loss Cost: *Prod. loss cost is $1,000/hr on critical belt.*	6,000	1,000	5,000
Total:	6,456	1,076	5,380
Downtime in Hours	6	1	5

Figure 2-9: Maintenance analysis example 5–faulty repair work.

The following example will help to illustrate the "guessing" procedure (see Figure 2-10):

> Let us assume that we are suspicious of excessive bearing failures on equipment XYZ, and, furthermore, we suspect that the oiling frequencies are incorrect. We proceed to "guess" at what would happen if we were to change these frequencies.

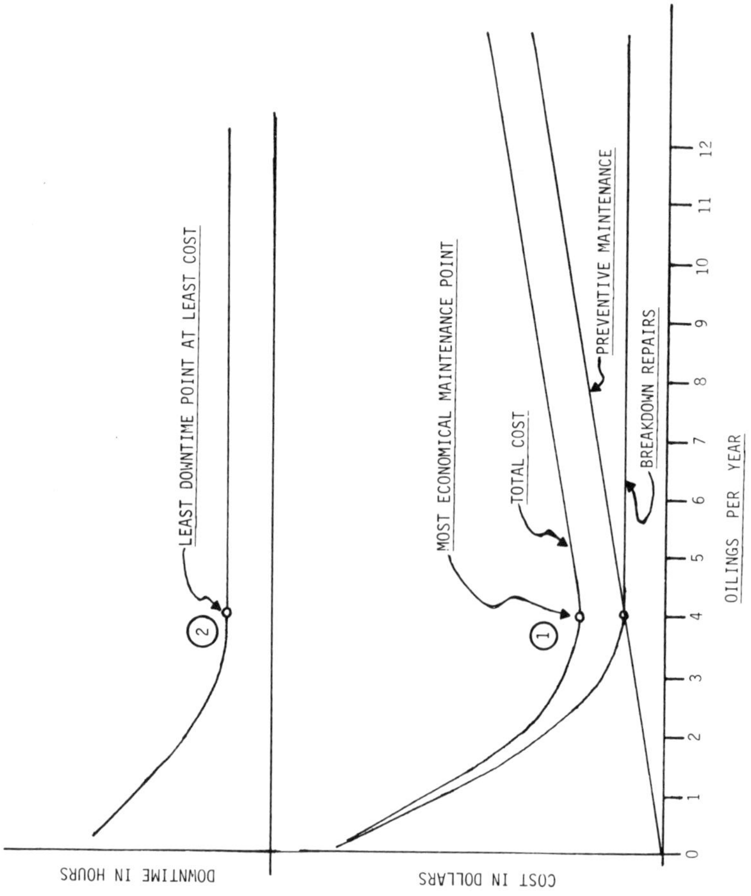

Figure 2-10: Maintenance analysis—guessing procedure.

Concerning preventive maintenance cost, we can assume that its cost will be a straight line function. In other words, every time we increase the frequency by one, the preventive maintenance cost will go up by one cost unit.

Breakdown repairs are, of course, astronomical with zero oilings. However, this cost drops sharply as oiling is started. We guess that bearing failures will be negligible when we reach a frequency of four times per year.

When the Preventive Maintenance Cost Curve and the Breakdown Cost Curve are added together, we get the most economical maintenance point (1). Point (1) is described in much more detail in the first few pages of Chapter 13.

We also conclude that the downtime hours will stop diminishing when the oiling frequencies go beyond points (1) and (2).

Corroboration of our "guesstimate" will take place by watching the equipment records. In this case, our "guesstimate" leads us to the conclusion that the bearings should be oiled four times per year.

The understanding of *how* to arrive at points (1) and (2) is *very* important since these points represent the two desired goals of a preventive maintenance system. The points are arrived at by *TRIAL AND ERROR*! At times, there is no other way.

Once again, it is important to understand that it is not necessary nor desirable to fill out the forms or plot the graph used in this section to illustrate the thought process. But, it is necessary to think in the manner described.

In order to further illustrate the maintenance analysis function the following case studies are offered to show the wide range of problems that one might face in the process of arriving at an intelligently designed maintenance program:

CASE STUDY 2D: The Cranes

At an aluminum extrusion plant which was touted as

being one of the largest in the world, I was investigating the PM program on the overhead cranes. There were 30 of these very large traveling cranes that spanned the width and the length of the huge main extrusion building. I found out that there were 4 electricians and 4 mechanics assigned full time to carry out the PM program for the cranes.

A thorough perusal of the instructions revealed that they were 100% inspections as opposed to positive work tasks (refer to Chapter 13 for an explanation of positive instructions). The cranes were working 24 hours per day, 7 days per week. The workers would start with crane #1 and proceed to #30 and then start all over again.

I then took a look at the inspection reports over a six month period to see what type of discrepancies they were finding. Not to my surprise, I found that the discrepancies were being repeated time after time each inspection round. I mentioned this to the inspectors and they said that they knew this was not the right way to handle maintenance, but they had no choice since the production department would never release the cranes as long as they were running. As a result the cranes were breaking down at frequent intervals.

The inspectors said that nevertheless they were making progress since the full round of inspections now took only 3 days whereas they used to take 5 days. This "improvement" was achieved by taking the previous week's report and marking it up in red. This took less time since very little changed from week to week!

I explained to management that the present method of maintaining the cranes was sheer insanity and was wasting 8 full-time men and achieving nothing. To add insult to injury the methods "improvement" mentioned above should have resulted in a reduction of men assigned to this useless task, but nobody took action to achieve this. The result was that the team had increased the frequency of the entire operation without authorization.

Management approved my recommendations for improvement which were as follows:

- o Rewrite the PM instructions to include as many "positive" as opposed to "inspection" types as possible drawing heavily upon the prior inspection reports.

- o Strike a "deal" with production to allow at least 50% of the average hours spent on breakdowns to date per week for accomplishment of positive PM instructions and to correct discrepancies found by PM inspectors with the caveat that this would continue for at least 3 months to prove or disapprove that the total downtime would drop. It was also agreed that they would return to the old system if no improvement could be shown. The present PM inspection team would be used to do both the positive instructions and the correcting of PM discrepancies.

The recommendations were implemented with a dramatic decrease in total downtime for PM and repairs to 1/3 of the previous total.

The moral to this case study is to be on guard for 100% inspection instructions that accomplish nothing!

CASE STUDY 2E: The Cooling Tower

This incident took place at Kennedy Space Center I believe on the eve of the first launch of the Apollo Spacecraft to the moon. I don't remember all the details of this incident which took place over 20 years ago, but I do remember the moral to the story.

At 1 A.M. I was awakened from a sound sleep to learn that the main cooling tower for the launch control complex was on fire! As I threw on my clothes in preparation for a dash to the center, I reflected on the facts related to the cooling tower.

It was a wooden tower and probably one of the largest in the world. I thought of the tons of water cascading down practically every inch of the tower and wondered at the improbability of it being on fire. I knew the problem it presented. The cooling tower was part of the air conditioning system for the launch complex computers and if it failed, the launch would have to be aborted. This was a particular problem that night since the base was full of dignitaries including hundreds of congressmen, senators, CEO's, generals, admirals and probably even a king and queen or two!

When I arrived, my launch support manager told me that the launch director wanted to have the fire chief arrested if he shut down the tower which he was threatening to do because he didn't want his men exposed to the electrical hazards of a wet tower on fire.

I climbed up the ladder to the top of the tower and saw that one of the four giant fans was stopped but that the other three were running. The fire was centered on the stopped fan and was burning its way downward into the tower. I quickly determined that the hazard to personnel was not that great and ordered the firemen to the roof to play their hoses directly on the fire. The tower continued to operate at 3/4 capacity which was adequate to prevent a launch abort.

Subsequent analysis determined the cause of the fire. An electrician had installed much too large overload elements in the motor starter. A short then caused wires to melt which dropped on the dry tower deck resulting in the fire.

A previous request for a redundant tower was turned down because of the huge expenditure involved. After the fire, a second *transite* tower was ordered indicating that the initial decision was incorrect.

The moral to this case study is that one should give careful and thorough consideration to the purchase of redundant equipment, particularly when a large production loss or loss of life is threatened by an equipment outage. In this case study, the lives of the three astronauts would have been placed in jeopardy by a crash launch abort which is what would have happened if the cooling tower failed completely.

I must tell you of a humorous incident that happened that very same night when I attended the morning stand up briefing. The launch support manager was going over the status of the various tasks (at launch minus 4 hours) when a message came in announcing the failure of a critical equipment item. I rushed to my car to drive to the location of the failure. The stand up briefing was also attended by all the aforementioned dignitaries (kings, queens, generals, congressmen, etc.) who also rushed to their vehicles.

When I arrived, there was a mob around the mechanic working on the failed equipment item. I pushed and shoved to get close and saw many heads squeezed close to the mechanic's head trying to get a good look at the problem area. I saw many generals' and admirals' stripes.

Finally the mechanic stood up and shouted, "Unless you S.O.B.'s get the hell out of here and give me some room to work, I'm leaving and *you* can fix it then!" Shortly thereafter, all of the dignitaries turned around and sheepishly left dragging their tails behind them.

The moral to this humorous incident is that we should remember the importance of the maintenance mechanic's skill and not to get too wrapped up in the importance of our paperwork and procedural systems. This point is emphasized in Chapter 2 under priority 1 which is to "assure the availability and competence of maintenance labor."

11

Analysis of Critical Problems

GENERAL

From time to time, problems occur that are of such a serious nature that a formal and immediate investigation is justified. Furthermore, analysis and corrective action performed at top management level is considered mandatory. Critical problems may fall within the following areas:

- o Maintenance failures causing serious production stoppage.
- o Operational mishaps causing serious production stoppage.
- o Maintenance failures or operational mishaps which *may* cause serious production stoppage.
- o Maintenance failures or operational mishaps which caused an accident or potential serious safety hazard.

The purpose of formalizing the investigation is to insure that the problem receives a thorough and penetrating analysis designed to prevent problem reoccurrence. Nothing is more embarrassing or inexcusable than to have repetition of a serious problem because the previous problem analysis was:

- o Not made
- o Sloppy

- o Incomplete
- o Offered no solution

The author has been repeatedly shocked as a maintenance consultant to find, in many of the plants that he surveyed, such a large number of critical events that were allowed to pass by without an investigation. As a matter of fact, many of these serious events occurred continuously over a period of many years. Plant personnel considered the event "just one of those problems that you had to put up with" in running their type of plant. In this section, a procedure to use in order to insure that critical problems are investigated is outlined, and then at the end, several examples are given out of the author's experience to illustrate the procedure. (The actual procedure may vary with each plant.)

REASONS FOR UNSATISFACTORY ANALYSIS OF CRITICAL PROBLEMS

Before a description of the required report is made, it may be helpful to describe some of the reasons why unsatisfactory analysis of problems is sometimes made. Practically all of these reasons can be traced to one person not assuming total responsibility for the entire analysis:

- o Production foreman ignores maintenance problem because it is not his responsibility.
- o Maintenance foreman ignores production problem because it is not his responsibility.
- o Mechanical foreman ignores electrical problem because he knows nothing about electrical work.
- o Person investigating problem takes someone else's word about some detail without making his own study because of lack of knowledge of subject matter or hesitance to cross organizational lines.
- o "It's just one of those things that

happen." "Why get upset about the inevitable?"

The point is that problems cross organization and technical expertise lines. It is up to the problem analyzer to cross these lines and make a thorough analysis regardless of his reporting position or technical skill. Furthermore, the problem isn't always a technical problem. It could, for example, be a personnel problem.

First line foremen are charged with the responsibility of promptly notifying management when critical problems occur.

CASE STUDY 2F: The Report

An incident occurred in my life that taught me a lesson in responsibility that I have never forgotten.

This incident occurred at Kennedy Space Center. The launch support manager called me at home at night to explain an equipment failure that delayed the launch countdown for one hour. He said not to worry about it because the equipment was fixed, running and the countdown had resumed. In accordance with our standard practice, he promised to have a full report on my desk by 8 A.M. Afterwards I was scheduled to brief the center director at 9 A.M.

At 8 A.M. I read the report which was not very clear. In questioning the launch support manager, who was accompanied by 5 other people, it soon became apparent that he had not visited the failed equipment personally since he was continuously referring the questions to others. I was unhappy about this and told him so. He said, "I am not an electrical engineer and since the failure was electrical, I didn't see the point in second guessing the electrical supervisor who knows much more about it than I do." I then tried unsuccessfully to squeeze the full story from the 6 people in my office. Each pointed the finger at someone else and said, "That's his responsibility." At this point I gave the launch support manager a lecture on the importance of *one* person taking charge of every job so that all the elements could be tied together. This philosophy is described in Chapter 15, "Daily Job Planning - Who is Responsible?" and

Chapter 16, "Project Supervisor Concept."

An hour later I briefed the center director Dr. Kurt Debus. I was feeling insecure and had a guilty feeling since I didn't feel I was well prepared. Dr. Debus, incidently, was second in command to Dr. Werhner von Braun in Germany and one of the top rocket scientists in the world that the U.S.A. hired to head up our space program after World War II. He was a dissatisfied with my explanation as I was with that of the launch support manager. He asked me to describe the failed part. I told him I could not because I had not seen the part. He asked me for pictures of the part. I told him I had no pictures. He then proceeded to berate me unmercifully for 1/2 hour for doing such a sloppy job of investigating the failure.

He explained what he wanted of me in the future as follows:

- It was up to me to have the wisdom to determine if my presence at the site of a failure was necessary. In the way of guidance he said that in the case of any incident causing a launch delay, it was mandatory that I be present since I was in charge of maintenance.
- In the case of this incident, I should have personally taken charge of the repair and subsequent investigation. Since I had failed to do my job properly, he was taking over responsibility for the continuing investigation. I was flabbergasted at the amount of detail he got into during the ensuing weeks to pinpoint the reason for failure and the final fix. Later I learned that it was this very thoroughness that had made him a success in Germany and that it was now a way of life with him.
- Since the countdown was delayed, it was mandatory that we find out the cause of the failure in order to do whatever was necessary to prevent it from happening again. In the case of this incident,

valuable information was lost by not taking pictures while the equipment was down since you could not do this after the equipment was fixed and closed up. Also, if the failed part was removed, pictures should be taken of it from several angles.

- I should include in my presentation to him at 9 A.M. an opinion of whether or not the fix was temporary or permanent. In the former case, he considered that we were in an emergency status since we did not know enough to fix it permanently and therefore the investigation to determine the permanent fix should proceed on an urgent basis.

The lesson of this incident applies equally well in every day manufacturing activities as I learned later. The case studies that follow show how the response to failures was not only equally sloppy, but that this type of response is more the norm than one might expect (in my experience).

DESIGNATION OF CRITICAL PROBLEMS

After being notified that an incident has occurred, the vice president of operations, or equivalent position, should determine whether or not to classify the incident as a critical problem.

RESPONSIBILITY FOR ANALYSIS

The maintenance manager, plant manager, or vice president of operations should designate the *single* person responsible for making out the Critical Problem Report. All reports, however, should be signed by one of the above three persons, indicating top management interest in having the problem solved. The writer of the report should assume *full* responsibility for the worth of the analysis and should make a presentation of the report to top management.

All throughout this book I emphasize the absolute necessity of having only one person responsible for *one project.* In the famous book entitled "In Search of Excellence" they say that it is characteristic of all successful projects that the project had a single leader who they called the "champion" of the project. Some of the situations where a single person only is needed to direct the project are:

- o Critical problem analysis
- o Project planning
- o Project execution

REPORT DUE DATE

The report should be submitted in stages as follows:

- o Verbal Report

  Due within the hour to the maintenance manager. The manager should verbally inform the plant manager and vice president of operations immediately thereafter.

- o Final Report

  A final written report is due on the desk of the maintenance manager, the plant manager and vice president of operations within 24 hours of the incident.

REPORT FORMAT

The report shall use the following format:

Critical Problem Report

No.________ Date________ By________

1. What happened?
2. What corrective action was taken?
3. What was the impact on production?

4. What was the cause?

5. What will be done to prevent recurrence?

The "number" listed above shall be an eight digit number determined as follows:

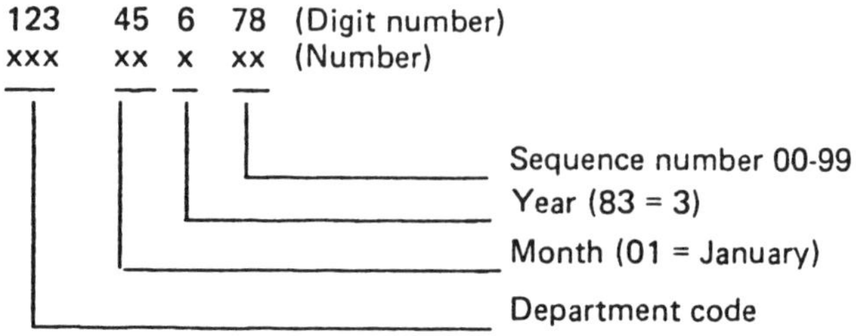

WHAT HAPPENED?

Explain what happened including what, when (date, hour) and how. Be complete, name who and what was involved.

WHAT WAS IMPACT ON PRODUCTION?

What was production downtime hours? Completely describe impact. What else was affected?

WHAT WAS CAUSE?

What was cause of problem?

- o Lack of preventive maintenance
- o Operator error
- o Operator carelessness
- o Faulty manufacture
- o Faulty installation
- o Act of God
- o Design error
- o Etc.

Go into sufficient detail to surface specific cause as opposed to general cause.

WHAT CORRECTIVE ACTION WAS TAKEN?

What was done to correct the problem? Was the fix temporary or permanent?

WHAT WILL BE DONE TO PREVENT RECURRENCE?

How will you guarantee that the event does not reoccur?

If you are not sure of how to prevent reoccurrence, state so and describe recommended investigation and research effort required to arrive at the final solution.

CRITICAL PROBLEM LOG

The secretary to the plant manager or vice president of operations should maintain status of all critical problems in a log book. Status shall be obtained weekly from the maintenance manager on unresolved problems.

A sample log book sheet is shown in Figure 2-11.

Critical Problem Log																		
Number	Date of Occurrence	Description	Status Code															
			DEPT 1				DEPT 2				DEPT 3				DEPT 4			
			1	2	3	4	1	2	3	4	1	2	3	4	1	2	3	4

Figure 2-11: Critical problem log.

The age of the problem will be shown in the log book. The secretary should list the problems in descending order of age. This will provide increasing pressure on the maintenance manager to resolve the problem as its age increases. As the problems are successfully resolved, they can be transferred to a "Resolved Critical Problem Log" which will be in a sense, a log of progress.

I suggest using only one of the following phrases in the status column:

Status	Code
Unresolved	1
Temporary fix made	2
Permanent fix design only complete	3
Permanent fix implemented	4

DISCUSSION MEETINGS

Meetings should be called by the plant manager or vice president of operations, as necessary, to update and satisfy him on the progress towards solutions to outstanding problems.

EXAMPLE

In order to illustrate the proper filling out of a critical problem report, the following example is offered:

CASE STUDY 2G: The Fuel Pump

Critical Problem Report

No. 511-113-04 Date 11/24/82 By John Jones

1. What Happened?

8 P.M. (Nov. 11)

Fractionating tower operator Smith reported over temperature condition on south main fuel pump bearing to control room.

9 P.M.

Operator Smith reports outage on main fuel pump.

9:10 P.M.

Auxiliary fuel pump started manually. Tower capacity reduced 30% to compensate for smaller size auxiliary fuel pump.

9:15 P.M.

Shift maintenance supervisor, Dave Doe, notified. Doe decides to have repair started on day shift due to existing heavy workload.

7 A.M (Nov. 12)

Maintenance manager notified of incident at home by midnight foreman.

7:15 A.M.

Maintenance manager notified plant manager of incident.

8 A.M.

Dismantling of damaged fuel pump initiated.

10 A.M.

Dismantling completed. Assessment of damage as follows:

South bearing seized, melted, totally destroyed

North bearing damaged

Shaft broken

Flexible coupling damaged

Motor tripped on overload. Motor OK

Impeller badly damaged

(Pictures of the damaged pump are attached herewith)

2. What Corrective Action Was Taken?

11 A.M.

Machining of new shaft started. Shaft completed 3 P.M.

12 Noon

Bearings confirmed as stock item. Item out of stock. Flexible coupling found in stock. New impeller located in stock.

3 P.M.

Purchase order complete. Bearings ordered air freight by phone.

9 A.M. (Nov. 13)

Bearings arrived in plant in receiving. Maintenance supervisor notified of delivery 10 A.M.

12 Noon (Nov. 13)

Pump rebuilt, reinstalled and put into service.

3. What Was Impact on Production?

Outage time:	9 PM November 11 to 12 Noon, November 13	39 hr
Production loss:	39 hr x $5,000/hr x 0.3 =	$58,000

4. What Was the Cause?

Bearing was a manually oiled type requiring daily oiling. Regular oiler off on vacation. Replacement oiler neglected to oil bearing. Bearing wasn't oiled for two weeks. Although there was a written oiling check sheet, foreman neglected to review sheet or give sheet to replacement oiler. Replacement oiler had filled in on this job several times previously and was thought to be familiar with assignment. Oiler interviewed and stated since all other similar pumps had greased bearings, he assumed this pump had greased bearings.

5. What Will Be Done to Prevent Reoccurrence?

This failure was unjustified. Furthermore, the length of the outage was found to be excessive as a result of a detailed analysis of the incident. The following is an analysis of the incident broken down by technical, preventive maintenance and communication aspects.

Technical Problem

The cost of this outage alone would justify a full-sized standby fuel pump electrically interlocked to start up upon failure of the main pump pressure.

The original design of the fuel pump calls for greased ball bearings. Someone made a design modification on this pump altering the bearing type. No record could be found of this modification or of previous bearing failures.

Furthermore, no maintenance supervisor recalls the modification. This mod is in violation of established procedures calling for the chief engineer to sign off on equipment mods.

Preventive Maintenance Problem

The oilers are in possession of daily check off sheets for the oiling route. Investigation of procedures reveals that the sheets haven't been used for at least two years. The present oiler has been on this assignment for one year. The maintenance foreman in charge of the oilers doesn't recall when the practice of using the sheets was discontinued. The maintenance superintendent says that he gave no order to discontinue use of the sheets and assumed they were being used.

Communication Problems

There were several serious violations of established communication procedures that resulted in excessive production lost time. These are:

- Failure of the operator or control room to report bearing overheat to the maintenance foreman responsible for oiling. This might have prevented the failure.
- The night shift foreman used poor judgment in not assessing this incident as a critical problem. He assumed that since another fuel pump was on the line that the problem could wait until the morning. The control room violated existing procedures in not notifying the

plant manager of a reduction in tower output.

- Concerning the location and delivery of replacement bearings--subsequent investigation revealed that suitable bearings were in stock listed under a different equipment item. The stores supervisor stated that he would have picked this up immediately if he were notified of the existence of a critical problem, which he wasn't.

  In addition there was no reasonable excuse for the ordering of bearings being delayed until 3 P.M.

  A further foul-up caused by lack of communication between maintenance and stores resulted in the delivered bearings sitting in receiving for one hour before notification to maintenance.

The problems with communication were attributed to:

- Violation of existing procedures.
- Poor judgment and lack of sensitivity to the criticality of the problem on the part of the several persons involved.

Recommendations

- That the auxiliary fuel pump be replaced with a full-sized unit, electrically interlocked.
- That the oilers' check-out procedure be reactivated after suitable reindoctrination of affected personnel.
- That the communication problems of this incident be reviewed with affected personnel together with the existing critical problem reporting procedure.
- That maintenance engineering look into the stocking problems surfaced by this incident.

- o That the advisability of disciplinary action be investigated further.
- o That modification procedures and failure feedback reporting procedures be reviewed with maintenance foremen because of the lack of records on an apparent previous bearing problem.

COMMENTS ON THE EXAMPLE

What frequently happens on critical problems is that the recording and analysis ends at step 3. Or even worse, the details of steps 1 through 3 are left out. The foreman in this case says "We fixed it; the incident is resolved." For example, the report might have been made verbally as follows:

1. What Happened?

The south main fuel pump broke down.

2. What Corrective Action Was Taken?

A bearing burned out tearing up shaft, flexible coupling and impeller. All parts were fixed and reinstalled by 12 noon on November 13th.

3. What Was Impact On Production?

39 hours downtime @ $5,000 per hour x .3 = $58,500

The point to be made is that when a serious event like this occurs, we are not only interested in eliminating future reoccurrences, but we are also interested in the details of *why* the downtime lasted for 39 hours and how we can cut the repair time to a bare minimum if the failure reoccurs.

Unfortunately, the shortened story told in the above example is frequently enough to satisfy some people. This story says *nothing* about the thousands of dollars wasted by a part sitting in receiving, or of a supervisor arbitrarily deciding to delay the rebuild, or in the foul-up on preventive maintenance instructions. These problems are only surfaced and resolved by the inquisitive mind and a thorough investigation.

RELATED CASE STUDIES

CASE STUDY 2H: Process Plant Bearing Problem

Critical Problem Report

No. 513-013-01 Date 1/24/80 By F. Herbaty

1.0 What Happened?

Production worker noticed southeast bend pulley bearing smoking on belt #3 at 7 P.M. on January-21, 1980. Bearing was immediately greased. By 10 P.M., it became apparent that the bearing was burning up and could not be saved.

2.0 What Corrective Action Was Taken?

At the change of shift at 11 P.M., 2 men were assigned to replace the bearing. A Rex P-2215 bearing available in stock was installed. Belt #3 was shut down from 11 P.M. to 3 A.M. on January-22 to accomplish this task.

The work was covered on WO #84866. The cause of failure noted on this work order was "lack of lubrication."

3.0 What Was the Impact on Production?

Belt #3 was down from 11 P.M. to 3 A.M., or four hours, for repairs. Since the silos were very low, this caused an almost immediate shutdown to both slurry mills, which were down until approximately 6:30 A.M. on January 22, 1974, or approximately seven hours. (One mill was down 4 1/2 hours, the other 7 hours.)

This outage did not immediately affect kiln production, but according to the superintendent, the decrease in slurry output *may* cut into kiln production at a later date due to low slurry tanks and numerous problems in slurry mill feed systems, including belt #7.

4.0 What Was the Cause of Failure?

4.1 General:

Investigation revealed numerous items which may have caused the failure. These items are discussed in the following paragraphs.

4.2 History:

Information in the work order files is not completely clear, however, it appears that there has been approximately eight bearing failures at the counterweight bend pulleys in the past fourteen months. 2,000 lbs. were added to the counterweight in 1976 and an unmeasured quantity approximately ten months ago.

According to the original design drawing (Hewitt Robin Sketch No. 1629), the original bearing specified was

"Dodge Type E Pillow Block Bearing"

The stock card for this bearing specifies a Rex P2215 as an equal to the original bearing. The original bearing, however, hasn't been ordered or stocked for the past seven years. Furthermore, the P2215 bearing is *not* equal to the originally specified bearing.

4.3 Bearing Seal:

On page 400 of the Rex Bearing catalog, different seals are rated as follows:

4--Excellent

3--Good

2--Fair

1--Limited

0--Not Acceptable

The existing situation at the bearings in question consists of an atmosphere of abrasive materials, frequently wet, causing a coating and buildup on the shaft. As a matter of fact, it is reported that these bearings are frequently submerged and running in this abrasive material.

The seal used on the Rex-P2215 bearing is a P seal which is practically the worst for the application having a 1 rating. The M seal has a 4 rating and is the one that should be used. This error is considered to be the major reason for failure.

4.4 Bearing Load

The counterweight load is the principal load on the bearing in question. Its exact weight is difficult to find, however, the following calculation is an approximation:

	Pounds
Original weight from equipment file in engineering	7,240
Counterweight added in 1976	2,000
Counterweight added in 1978 (estimated at 20%)	2,000
Estimated build-up weight (from engineering files)	2,800
Subtotal	14,040
Add 50% for heavy service (see page 10-44 of Dodge Selection Guide Book)	7,020
	21,060

$$\text{Load per bearing (4 bearings)} = \frac{21{,}060}{4} = 5{,}265$$

$$\text{Vector component load } 1.5 \times 5{,}265 = 7{,}898$$

The allowable load of the bearing in use (Rex P-2215) is 7,175 lbs. In other words, it is too light.

It is important to note that the maintenance foreman over belt #3 states that the last time the bearing failed, the counterweight was *completely submerged* in stone and that it was cocked sideways a couple of feet thus binding it in its rails! This condition would place practically an infinite load on the bearing.

4.5 Bend Pulley Scrapers:

Presently there are scrapers on both bend pulleys which cascade product directly on top of the counterweight. In addition, there is spillage directly from the belt which dumps product on top of the counterweight and the bearings.

The original design attempts to take care of this situation by having a hole directly below the counterweight which leads into a chute which discharges the material onto belt #7. This design is unsatisfactory for the following reasons:

- o It does not prevent buildup on top of the counterweight.
- o The hole below the counterweight frequently plugs up.
- o The chute leading to belt #7 is unsatisfactory since it allows spillage to accumulate on both sides of belt #7.

4.6 Bearing Misalignment:

According to the maintenance foreman, it is not unusual to cock the shaft as necessary to properly center the belt when reinstalling a bearing. In other words, there is need for a bearing with sizeable misalignment capabilities such as is inherent in the Rex type design.

4.7 Lubrication:

The work order states that the bearing failed due to lack of lubrication. A check with the oiling records shows that the "Discharge End of Belt #3" was lubricated on January 21, 1974, the date of the incident. A further check shows the same notation a week previous.

Questioning of the oiler involved confirms that the quoted statement *includes* the bearing that failed and that he did in fact oil the bearing on that date.

Presently, there is *no* written check sheet ordering the lubrication of the subject bearing. Instead, there is a verbal agreement to lubricate it once per week. This agreement is unsatisfactory since it is also considered proper to drop the lubrication if more pressing work presents itself. Basically, there is *no way* to satisfactorily audit compliance to this type of instruction or to confirm that the oiler is telling the truth.

The existing lube frequency of once per week appears to be too frequent and is probably required because of the incorrectly selected seal. When a bearing seal of the proper type is installed, it is recommended that the lube frequency be reduced to once per quarter.

5.0 What Should Be Done to Prevent Recurrence?

5.1 Bearing Design:

The excessive bearing failures can be eliminated by proper bearing selection as follows:

- Radial Load Capacity - 10,550 lbs. at 100 RPM See Rex Catalog, Page 399, for 9000 Series or equal.
- Seal Type M or equal. See Rex Catalog, Page 400.
- Fully Self-Aligning Bearing required similar to Rex Design.

It is recommended that Rex MPA-215F (M Seal) be used.

It is furthermore recommended that a 2' x 1' hole be cut into the grating directly below the bearings to prevent buildup on the bearings. This hole can be partially closed by installing 1/8" x 1" bars on 3" spacing for safety purposes. Consideration may be given to installing a sheet metal cover over the bearing in addition to protect it from buildup.

5.2 Lubrication:

Initiation of the preventive maintenance program specified in the Manual will provide for specific "engineered" instructions *and* a basis for auditing compliance with the instructions.

5.3 Similar Bearings:

Engineering should make a study to determine

which other bearings in the plant fall under the same circumstance as this bearing. These bearings should then be programmed for similar corrective action.

5.4 Product Spillage:

Product spillage problems surfaced in this report should be made the subject of a separate study by engineering.

5.5 Analysis of Critical Problem:

The problem with the subject bearing is a serious one of relatively long standing. The fact that an investigation has not taken place automatically some time ago is a source of concern.

It is recommended that all production, engineering and maintenance personnel be indoctrinated in the purpose and details of the procedure for analyzing critical problems.

5.6 Bearing Replacement Method:

The bearing replacement time for this incident was four hours. Some foremen say that the job can be done in one and one-half hours. Since downtime is of major concern on this belt, it is recommended that a job method study be made to determine if there is any way to cut down on the replacement time.

The bearing in this case was burned out with a torch. This gives rise to the question of whether or not the shaft has been damaged to the extent that it is causing bearing failures.

5.7 Head Pulley Slippage:

It has been reported that despite the addition of counterweights two times that slippage is still experienced at the head pulley. This is worthy of a separate investigation.

5.8 Stocking Problems:

Someone, seven years ago, specified an "or equal" bearing that was not equal to the original design on several counts. Of course, the impact of this error is less because neither bearing is suitable for the job.

This occurrence should be used to illustrate to people making the "or equal" decision to be careful in the process and to make the necessary evaluation as a prerequisite to selecting an "or equal" item.

5.9 Counterweight Buildup:

Production should be instructed never to let counterweight buildup exceed two feet.

CASE STUDY 2I : Process Plant Critical Conveyor Problem

The problem illustrated by this example was not a single incident but a recurring problem over a period of 7 years and, therefore, does not exactly follow the format previously described. In this case, management decided that the recurring problem was serious enough to warrant a special study.

1. General

Belt #7 is a critical belt since all product flows through the belt and there is no redundant belt.

An analysis of work orders and the log book for belt #7 reveals the following highly repetitive problems:

a. Repairs to belt - tears and clips and saddles

b. Buildup at tailpulley

c. Buildup at headpulley

d. Renewal of return rollers

e. Cleaning of dirt from rollers

f. Bearing failures

g. Dynamatic problems

h. Coupling problems

2. Product Spillage

One of the biggest causes of downtime results from product spillage at the tail end. Product is spilling out of worn skirts and holes in chutes. It is believe that this spillage is the primary cause for problems a, b, d, e and f.

It is firmly believed that performance of the tasks described in the attached preventive maintenance instruction will virtually eliminate these problems.

In order for this preventive maintenance instruction to be effective, belt #7 must be put into proper condition by doing the following:

- o Renew skirts.
- o Reline chutes on brushes and wobbler discharges.
- o Countersink clips and eventually replace with vulcanized splices and repairs.
- o Reinstall scrapers.
- o Close opening in table top above counterweight.
- o After all of the above items have been taken care of; the tail end area should be thoroughly cleaned.

A special problem exists with product spillage from belt #3 to belt #7. It is recommended that this be the subject of a separate study by engineering.

Another major spillage problem occurs at the head end. For the period April 11, 1973 to January 29, 1974, there were forty-three downtime occurrences to dig out product accumulation under the return strand at the head end. This problem is caused by material being carried back on the return strand. The problem is aggravated by a relatively small clearance between the return strand and the floor plate. There are several possible ways to eliminate this problem:

- o Countersink belt clips and/or

install vulcanized splices and repairs. This will permit installation of belt scrappers and/or brushes.

- o Modify the shape of the floor plate at the accumulation to prevent buildup.

It is recommended that engineering study this problem and come up with the most advantageous design.

3. Bearings

There have been nine (9) pulley bearing failures in the past fourteen (14) months. The original design on these bearing is as follows:

"Rex ZPA-315-F"

This bearing is satisfactory from load and misalignment standpoints but has an unsatisfactory seal. The seal for this very severe abrasive dust condition should be a type M seal.

Approximately two (2) years ago, the purchase of this bearing was discontinued and the following submitted:

"Rex P-2315F"

This bearing is even more unsatisfactory since its allowable load is only one-half of the original design.

Furthermore, it was observed that some of the installed bearings were of the 2-hole type instead of the original 4-hole type, and were mounted on unsatisfactorily designed bases.

It was noticed that some of the 4-hole installed bearing were mounted on 2-hole bases with inadequate base support area.

In addition, several of the bearings were submerged in product spillage to the point where the shaft and the bearing were being worn.

Also, connections were made to grease points on bearings located on the opposite side of the catwalk with inadequate rubber hose which was fastened to pipe nipples with twisted wire.

It was also observed that there is a high rate of bearing failures on the return and troughing idlers. It is believed that this is caused by an insufficient frequency of bearing lubrication. (This frequency was increased from yearly to semi-yearly in the attached preventive maintenance instruction.)

It is recommended that engineering make a complete study of *all* bearings on belt #7 and come up with the proper bearing design. The study should include the following:

- Load calculations and requirements
- Seal specs (M type recommended)
- Misalignment specs (fully misaligned capability type recommended)
- Bearing base design (adjustable type recommended)
- Proper design of lube point extensions

In order to alleviate the problem of spillage accumulation around bearings, it is recommended that:

- A portion of decking be removed around bearing base so that accumulation can fall through.
- That a metal shield be installed over bearings.

4. Dynamatic

Four (4) failures of the dynamatic have been experienced in the past fourteen (14) months. This has been a consistent problem in the past years.

The impact of a dynamatic failure is very severe because, when the dynamatic fails (since there is no spare), it is replaced by a solid jack shaft. This causes approximately four (4) hours of unscheduled downtime. The result of this change is that there is an instantaneous shock load on the motor, couplings, gear box and the entire conveyor systems, which has caused:

- Coupling failures between motor and gear box
- Motor failure and motor shaft twisting
- Belt splice damage
- Excessive belt strain
- Causes top strand to rise off idlers during startup, hitting magnet

An investigation reveals that all dynamatic failures have been in the bearings. Within the past month, a redesigned dynamatic has been installed. The redesign includes the addition of two (2) extra bearings. It is hoped that this modification will improve the situation. An investigation of the possible causes of bearing failure reveals the following:

o According to the Dynamatic Company, improper alignment procedures are a major cause of bearing failures. It is, therefore, recommended that correct alignment be accomplished with the use of dial indicators.

o It has been reported by the maintenance foreman that the internal components of the dynamatic have been so heavily encrusted by dirt that it takes two (2) days to scrape the parts clean. This situation very likely causes an unbalanced condition and vibration and contributes to bearing failures. In order to alleviate this condition, it is recommended that a duct be installed through the nearby roof, complete with filters and a gooseneck to lead clean air into the dynamatic air intake, as opposed to the present air which is saturated with erosive rock dust.

o An investigation of the causes of bearing failure reveals that the present type of grease has a tendency to channel away from the bearing races. Mobil Oil Company has recommended that a new and improved product be used. This is Mobile Grease #27. This product is currently being tested in a smaller dynamatic.

o It has been reported on numerous occasions that the motor and dynamatic base bolts have loosened allowing misalignment. In the past when this has occurred, the bolts have been tightened without going through realignment procedure. It is recommended that realignment be accomplished whenever this occurs and that the bolts be retightened and torqued to their maximum specifications and that the motor, dynamatic and gear box be doweled to base.

In addition, a weekly inspection has been specified of those bolts in the attached preventive maintenance instruction. (See Figure 2-12)

Belt 7: Summary of Recommendations

Item No.	Description	Responsibility
1	Implement PMI	Mechanic maintenance
2	Replace chute liners	Mechanic maintenance
3	Countersink clips and reinstall scrapers	Mechanic maintenance
4	Close opening in table top over counterweight	Mechanic maintenance
5	Thoroughly clean tail end area	Operational maintenance
6	Investigate spillage from Belt 3 to 7 and recommend design change	Engineering
7	Investigate spillage at headend and recommend design change	Engineering
8	Investigate bearing design, recommend new design, implement new design, correct stock problems	Engineering
9	Follow-up on success of bearing redesign on dynamatic	Mechanic maintenance
10	Use correct realignment procedures on dynamatic and motor (check present alignment and dowel to base)	Mechanic maintenance
11	Install dynamatic clean air duct	Mechanic maintenance
12	Implement new dynamatic grease	Mechanic maintenance

Sheet 1 of 1

330207 MACHINE NUMBER	P. M. INSTRUCTION	WORK ORDER NO.

MACHINE DESCRIPTION Belt 7 COST CENTER NO. 330

		MECHANICAL					ELECTRICAL					LUBRICATION				
		W	M	Q	S	A	W	M	Q	S	A	W	M	Q	S	A
1	Replace all skirts on tail end (10")			8												
2.	Adjust Skirts and Scrapers	2														
3.	Inspect Crusher Stone Box and Chute Liners. Replace as Necessary.		¼													
4.	Starting at Head Pulley and Proceeding to Tail Pulley, Inspect the Following: • Check Tightness if Motor, Dynamatic and Reducer Base Bolts for Tightness. • Check Head Tail, Counterweight, and Bend Pulley Bearings. • Check all Top and Return Idlers • Check for Excessive Buildup of Any Type	1														
5.	Check Belt Running While Empty	½														
6.	Check Safety Pull Switches			½												
7.	Lubricate Motor Bearings								½							
8.	Inspect and Clean Motor Starter										2					
9.	Lubricate Pulley Bearings											1				
10.	Lubricate Dynamatic													½		
11	Lubricate Troughing and Return Idlers														32	
	TOTALS	3½	¼	8½					½		2	1		½	32	

Figure 2-12: PM instruction.

Why was this problem allowed to continue over a 7 year period? Would this have continued to happen if the workers owned the plant?

CASE STUDY 2J: The Handrail Welders

This story took place in a process plant. There was a line of huge smelters, many stories high on one wall of the smelter building and a line of Bessemer Converters on the opposite wall. The two groups of equipment were separated by a narrow crane aisle.

The traveling crane, which transversed the length of the building, was located at the underside of the roof about seven stories high. The crane carried a 20 ton pot, used to transfer molten ore from the smelter to the Bessemer Converters for further refining of the ore.

While studying the maintenance organization chart, it was discovered that there was a group of fifteen (15) welders whose sole job was to repair handrails on platforms and structural steel members of the smelters and converters which were being smashed by the ore pot hanging from the crane as it worked the aisle.

When asked why nobody stopped this seemingly unnecessary destruction of facilities, the plant personnel looked upon the author as a wise guy and invited me to solve the problem.

I had just talked the president into starting the Analysis of Critical Problem procedure and was anxious to find a good example for the first analysis.

I was shocked to discover that the fifteen (15) welders had been performing this work for ten (10) years. I needn't have been shocked because over the course of several years, I was to discover that many plants suffered serious problems that were reoccurring over many years. The personnel had come to look upon the critical problem as just part of the job of running the plant, in a manner similar to the need to periodically change spark plugs.

After arranging for a meeting of all top management and explaining that having fifteen (15) welders repairing smashed handrails for ten (10) years was not like changing spark plugs. It was a *major critical problem* that should be stopped immediately. I recounted my experience at other plants where critical problems were allowed to happen only once! I expressed confidence that the problem could be eliminated by myself with their help and cooperation. I immediately logged the problem.

I decided to visit and ride with the crane operator. I made it to the crane level at the underside of the roof and almost immediately fled in sheer panic unable to breathe. The underside of the roof was filled with poisonous fumes from the smelter and converters and the temperature must have been 110°F. I returned later with oxygen breathing apparatus. I could not understand how the crane operator could stand it. He said you get used to it and jobs were hard to find and that is why he stayed.

Investigation soon revealed that all twenty (20) roof ventilator motors were out of commission and were operating on natural draft only. The following was immediately recommended and approval to implement was obtained.

- o Repair all roof ventilator motors.
- o Sealing of leaks on smelters and converters.
- o Enclosing and air conditioning of crane cab and providing electrostatic air cleaners.
- o Provide oxygen breathing apparatus in crane cab for emergencies.
- o Providing of windshield wipers and washers for crane cap operator so that he had a fighting chance of seeing what he was doing.

Meanwhile, I found out that many of the production foremen had added catwalks that stuck out in the crane aisle so they could see what was going on. All of these were removed after determining they were not necessary. The crane operator had a tough time dodging all these catwalks.

The handrail stanchions were redesigned, putting springs in the handrail sockets so that if the handrail was hit by the 20 ton crane pot, it merely bent over and returned after the crane had passed.

All of these action items reduced the problem to a minor manageable one. The question is why was this allowed to continue for so many years and would it have happened if the workers owned the plant? I think the answer to this question is clearly no, pointing out that the root problem is related to the interaction between management and labor.

CASE STUDY 2K: The Screw Conveyor Bearings

The author was asked to look into the failure of several hundred bearings per year on the screw conveyor receiving particulate matter from a battery of huge electrostatic precipitators in a large process plant. These failures had been occurring since the plant was started ten (10) years previously.

After interviewing the maintenance personnel familiar with the problem, it was discovered that the bearing was inside the screw conveyor itself and was, therefore, inundated in the material being conveyed and was subjected to a temperature of 350°F.

After deciding to spend a few hours looking into the file on the ESP's, I soon found a letter written by the chief engineer of the firm that designed the plant, (about ten years previously) to the company that was to provide the screw conveyor. The chief engineer gave the specifications on the bearings he planned to use on the screw conveyor. The return letter stated that the bearings anticipated for use had seals that were good for only 250°F and would fail under the design temperature of 350°F. Guess what? The process plant was using and stocking the bearings with 250°F seals. The design firm never followed up on the letter from the supplier and nobody in the process plant had enough intellectual curiosity to read the files where the answer to the problem was waiting for anyone interested enough to look.

Why was this problem allowed to perpetuate over so many years?

CASE STUDY 2L: The Ore Belt Cutting

It was the first day of my assignment to make a maintenance study at a large process plant and I was being conducted on a tour of the facilities. The guide was pointing out one of the very large ore carrying belts. There were six of these expensive belts all being fed by overhead hoppers by gravity.

As I watched, I noticed a large tear developing in the belt as if some giant knife were cutting it. I immediately showed this to my guide who pushed an emergency stop button.

The next day I came back to the site to find out what was causing the tear. I walked up to the level above and talked to the operator where the six hoppers were being fed by overhead cranes. The operator explained that it was his job to keep the hoppers flowing freely. This was a problem, since the ore would periodically cake-up and refuse to flow down by gravity. The tool he used to prevent this was a compressed air lance. It consisted of a 15 foot long, 1 inch diameter standard galvanized pipe with shut-off valve fed with 100 PSI compressed air through a long hose. The hose permitted him to service all six hoppers. When one of the hoppers started to clog up, he stuck the pipe into the inverse pyramid shaped hopper, turned on the compressed air and wiggled it around until the ore ran freely again. Asked if they had found the reason for the belt tear from the previous day, the operator said that the end of his pipe lance had broken off and flowed down into the horizontal belt where it jammed against the sides of the hopper. The broken end of the pipe then acted like a knife, cutting the belt. When asked how often this happened, he said it happened about twice per week and had been going on for seven years.

Furthermore, the operator stated that he had reported the incidents and the cause to maintenance dozens of times and finally stopped since it became apparent that no one was going to do anything about it. Why was this allowed to happen?

If I hadn't seen many things like this before, I would have been shocked. That day I recommended that vibrators be installed on the sides of the hopper. This solved the problem.

CASE STUDY 2M: The 1000 HP Motor Fire

As usual when starting the day, I checked with the chief-planner scheduler to see if anything unusual had happened during the night. He said, "Yes, a 1000 HP synchronous motor had caught on fire and was completely destroyed." A spare motor had been installed and was already in operation.

I checked with the maintenance foreman to find out what the cause of the fire was. He said they were unable to determine the cause and that men were now examining the burned out motor.

I immediately called the general manager and recommended that the motor be shut down since the cause of the fire had not been determined, and there was a possibility that the new motor would burn up for the same reason. Furthermore, there were 20 identical units and therefore shutting down one would not be a problem. He agreed and had the motor shut down.

It was later determined that the fire was caused by a faulty protection device. This device was repaired and the motor was then returned to operation.

Examples 2B through 2E illustrate the consequences of not investigating a serious incident or of an inadequate investigation. It also demonstrates that it is not rare for personnel to consider even major failures as commonplace events that "one must expect in our business" and feel no obligation to determine the cause and eliminate recurrences.

The most important point of this section is to differentiate between run of the mill failures which should be handled as described in Chapter 10, Failure Analysis, and major failures or catastrophes, which must be investigated immediately as described in Chapter 11, Analysis of Critical Problems, to insure that they do not reoccur.

Any big problem which continues to occur frequently over many years is a critical problem.

12

The Principle of Simultaneous Downtime

The Principle of Simultaneous Downtime (POSD), a concept conceived by the author, is particularly important in maintaining large, complex, expensive equipment items with high downtime costs.

The principle is quite simple--component failures should be synchronized to occur simultaneously, thereby limiting breakdown to only one occurrence instead of several.

In the ideal case, part A and part B each have a failure period of exactly 6 months. Downtime caused by failure of each part lasts 8 hours. Annual downtime caused by the two parts failing at different times will be: 2 parts x 8 hours per failure x 2 failures per year = 32 hours.

Replacing both parts when one fails will synchronize the failures so they will occur simultaneously. Annual downtime on the equipment can then be reduced from 32 to 16 hours.

However, in practice the concept is more complicated.

For example, consider the case of a 30-foot-diameter, 300-feet-long rotary kiln, installed in a process plant running 24 hours a day, 7 days a week. The material produced is in short supply and all product can be sold. Downtime on the kiln, therefore, causes an irretrievable loss in sales.

Downtime usually lasts two weeks, and the downtime cost is $10,000 an hour. Total cost for each downtime occurrence is: $10,000 x 24 x 14 = $3,360,000.

There are several reasons for the long average down time. These relate principally to the equipment's large size, for example: cool down and heat up times and brick replacement.

The loss in sales dollars when the kiln breaks down is tremendous.

A thorough failure-recording system will pinpoint which parts have failed in the past and how much downtime the failures have caused.

The maintenance manager must rely on experience in choosing the best maintenance approach to increase kiln productivity and reduce downtime.

In this case, he recognized that brick repair was the major cause of downtime and had replaced the original brick and insulation with different types during repair to try to determine which type was the best.

He was also familiar with the many problems common to the combustion system, the precipitators, and the input/output mechanical systems.

His first approach at reducing downtime was:

- Use as many repairmen as necessary (the entire maintenance department and some production people, if required) to repair the kiln as fast as possible.
- Maintain an adequate stock of spares.
- Make only those repairs necessary to return the kiln to operation as soon as possible.

With this approach, kiln productivity was 70 percent. The kiln was breaking down frequently for various reasons. Although the average length of the downtime period was reasonably small, the number of occurrences was high. And, the plant manager was unhappy because kiln productivity percentages in several other plants were higher than 70 percent.

The maintenance manager chose another approach. He noticed that brick failure would occur adjacent to repairs of previous failures. Repairmen were instructed to examine the condition of the brick adjacent to the area of failure and, if warranted by the inspection, renew an entire band of brick, perhaps 5, 10, or 15 feet wide, instead of replacing only the failed brick. Records were kept of which areas were renewed and the dates of renewal to preclude renewing brick unnecessarily.

A mechanical maintenance man, repairing a failed bearing on a primary air fan for the kiln's main burner, suggested to the maintenance manager that the bearing could easily have been replaced during the last downtime occurrence. The maintenance manager then ordered that whenever the kiln was down for repairs, a team of experts would thoroughly inspect the brick, the insulation, and the mechanical, electrical, electronic, and structural systems to determine if any components were likely to fail in the near future. The team would make recommendations for replacing or overhauling weak parts. If the repairs could be made within the time required to repair the part causing the downtime, then these repairs would also be made. In addition, the economics of replacing or repairing parts with longer repair times than that of the original failed part would be evaluated and appropriate orders issued.

Within 6 months, this revised approach increased kiln productivity from 70 percent to 80 percent.

At this point, the maintenance manager was concerned with further increasing kiln productivity. In annual dollars the remaining 20 percent represented approximately: 0.20 x 8760 hours per year x $10,000 lost production cost per hour = $17,520,000. Some of this money could be saved by using POSD.

The basic procedure for applying the principle of simultaneous downtime is:

1. Breakdown repair orders should be sorted according to failure items.

2. Records of each type of failure should be arranged chronologically.

3. Times between failures for each type of occurrence should be calculated.

4. Average times, and the longest and shortest times between failures should be calculated.

5. Results should be tabulated as shown in Figure 2-13.

6. Data should be analyzed so decisions about synchronizing failures can be made.

Figure 2-13 reveals that failures of kiln components 2, 3, 4, 5, 6, 7, 8 and 11 should be synchronized because:

- o Variations between the high, average and low times between failures are very small.
- o Ratios of downtime costs to repair costs are very high, ranging from 53.3 to 6.7. For example, cost of downtime caused by item 8 is 53.3 times that of the actual repair.

Item 1 was excluded from further consideration for two reasons:

- o Time between failures was too variable.
- o Ratio of downtime to repair costs was too low and the repair cost was very high. This brick area should be replaced only in the event of a failure.

The importance of the ratio of downtime to repair costs cannot be overemphasized. Consider item 8, for example, whose ratio is 53.3. In this case, an expenditure of $1,500 can save a downtime expense of $80,000. In cases such as this, we must also consider the quality of the repair part. For example, if the cheapest part can be purchased for $1 and the most expensive for $5 (assuming a commensurate increase in life expectancy and quality), then we should

No.	Component	Time Between Failures Months: High	Average	Low	Downtime for Repair, Hours	Cost X: Average Downtime Cost, Dollars	Cost Y: Average Repair Cost, Dollars	Ratio: Cost X/ Cost Y
1	Brick Area A	10	4	2	336	3,360,000	1,000,000	3.4
2	Brick Area B	4	4	4	280	2,800,000	400,000	7.0
3	Brick Area C	4	4	4	200	2,000,000	300,000	6.7
4	Insulation Area A	4	4	4	150	1,500,000	100,000	15.0
5	Insulation Area B	4.1	4	4	125	1,250,000	70,000	17.9
6	Insulation Area C	4	4	4	80	800,000	60,000	13.3
7	Burner Part A	4.2	4.1	4.1	16	160,000	4,000	40
8	Burner Part B	4	4	4	8	80,000	1,500	53.3
9	Burner Part C	8	6	5	4	40,000	10,000	4.0
10	Cooler Part A	12	7	3	40	400,000	40,000	10.0
11	Cooler Part B	4.5	4.4	4.3	30	300,000	10,000	30.0
12	Cooler Part C	17	11	6	10	100,000	8,000	12.5

Note: The above data is purely hypothetical and not related to actual numbers.

Figure 2-13: Principle of simultaneous downtime chart.

purchase the most expensive part since the increased cost is insignificant compared to the potential savings. Also, only entrust the "or equal" buying decision to someone who knows what he is doing and who has all the facts. Better yet, specify "no substitutions." There are several case studies in the previous chapter that describe how "or equal" decisions fouled up the works.

The decision to synchronize failures is difficult because of the variations in time between failures of many components. Careful study of the time between failures, downtime cost, repair cost, and the ratio between these costs is necessary.

Another consideration is whether a component failure will cause catastrophic failures of other parts and create safety hazards. If it will, the repair should be made during a planned shutdown. The repair should be timed to occur before the mean time between failures, and instrumentation should be installed to give advance warning of failure, if possible.

Application of the POSD approach in this case would have increased kiln productivity well above the 80 percent mark. In addition to the POSD approach on large complex equipment, it goes without saying that a sound preventive maintenance program should be in effect for all auxiliary or redundant equipment items that can be shut down for preventive maintenance without shutting down the main system.

POSD could also be used as the basis for scheduling planned periodic shutdowns for repairs. The frequency and duration of downtime, as well as the items to be repaired, can be determined by POSD.

A variation on the concept of periodic overhauls is the annual total plant shutdown, during which *everything* that needs it, is overhauled or repaired. This approach violates the POSD, since components are overhauled regardless of time between failures. This approach, however, may be justified in some cases where there is a seasonal drop in demand for the plant product close to zero demand. Another reason might be that the raw material feeding the plant is only available during a part of the year.

The annual shutdown tends to be favored in plants that run around the clock or where some aspect of the process and plant design favors it.

Another alternate that I experienced was in an aluminum extrusion plant where the aluminum furnaces were allowed to run around the clock until they literally collapsed, at the point they were completely rebuilt! (Just try and visualize everyone screaming at the top of their lungs as the molten aluminum comes spilling down the main aisle, "*Run for your lives!*")

ANALYSIS FOR ZERO DOWNTIME

Another special case of analysis is when we want zero downtime. In this case the equipment doesn't have to be large or complex, it could be any size. Some of the things that can be done to achieve zero downtime are:

- Use redundant electrically interlocked equipment so that the redundant unit starts up automatically when the other breaks down. In this case design of the interlocking circuit may be critical if the allowable outage time is small as is the case, for example, in the power supply to some computers where allowable outage time may be only a fraction of a second.
- Employ beefed up, conservative design.
- Renew components before expected failure at very conservative intervals, unless component is very costly.
- Prevent operational abuse through training and licensing of operators.
- Utilize operational monitoring (predictive maintenance) instrumentation to forewarn you of pending failure.
- Write conservative PM instructions.
- When components are replaced, use the best, most reliable part particularly when

the difference in cost between the best and the worst is minuscule compared to potential production loss cost.

- When failures occur, make an exhaustive analysis and redesign parts or systems to preclude failure by:

  - Destructive testing and analysis
  - Redesign for increased reliability
  - Beefed up PM
  - Reanalyze spare parts
  - Examine ratio of production loss to repair cost

  For example, when TWA overhauls its planes, they say that the plane is more reliable than when new because they incorporate many design improvements.

- Another example is spacecraft. NASA's philosophy on failure is that only one is allowed because of the possible tragic consequences of a single failure. They want to do everything possible to prevent a recurrence.

In summary, these are the special things one must do to achieve zero downtime.

13

Preventive Maintenance (PM)

GENERAL

Today's technology is advancing at an incredible rate. Plants are becoming ever more automated, resulting in an increase in the ratio of maintenance to production workers and job complexity. In the plant of tomorrow, the majority of workers will be those highly skilled in maintenance. Therefore, careful study to improve methods and reduce maintenance costs is imperative. Achieving these objectives will help save billions of maintenance dollars.

Specifically, such a study should examine methods used by industry to write preventive maintenance instructions. Although much has been published about selling, managing, scheduling, setting up paperwork systems, etc., little has been written about the most important part of the preventive maintenance program--the quality of the preventive maintenance instruction itself.

Finding personnel with all of the qualifications necessary to the writing of good preventive maintenance instructions is always a problem, and the smaller the number of people authorized to do so, the more the problem is compounded since in that case, each preventive maintenance writer must cover a broader area. If a shortage exists, it should be mitigated by outside or in plant training. In any event, management should put its most qualified people on the task of writing preventive maintenance instructions.

One of the most widely used reference sources for writing preventive maintenance instructions is the manufacturer's maintenance instruction book.

It is important when consulting this manual that the manufacturer's philosophy concerning maintenance costs be understood. First of all, the manufacturer designs a machine to be as reliable and maintenance free as manufacturing cost and competition will permit. Once the design is fixed, he may then tend to overwrite preventive maintenance instructions to, as nearly as possible, approach the zero breakdown level. He does this in order to achieve a reputation of reliability for his machine. He may be more concerned about the machine's reputation than he is about costs. Of course, the owner is interested in both. Furthermore, it would be difficult for the manufacturer to cover all the variable conditions his machine may be subjected to in one set of instructions.

Consider the example of the automobile manufacturer who recommends that the oil be changed every 2,000 miles. What would be the change in the car's total maintenance cost and reliability if oil changes were recommended every 4,000, 6,000, 8,000 or even 12,000 miles? Negligible, most likely, if the owner replaced the car every two years.

The difference between changing the oil every 2,000 miles versus every 12,000 miles represents a cost multiplier of six. This single cost may seem insignificant but, if applied to a plant with 100,000 units of equipment on which excessive preventive maintenance instructions are written, the total extra maintenance cost could be in the millions of dollars.

What can be done to reduce these total costs? Let's focus on a single piece of equipment and plot curves to keep track of the variation in maintenance costs and downtime percentage to supplement an analysis of breakdowns and the writing of maintenance instructions to prevent these breakdowns. Figure 2-14 shows three curves plotted against preventive maintenance activity level, i.e., downtime percentage, preventive maintenance cost and unplanned maintenance cost.

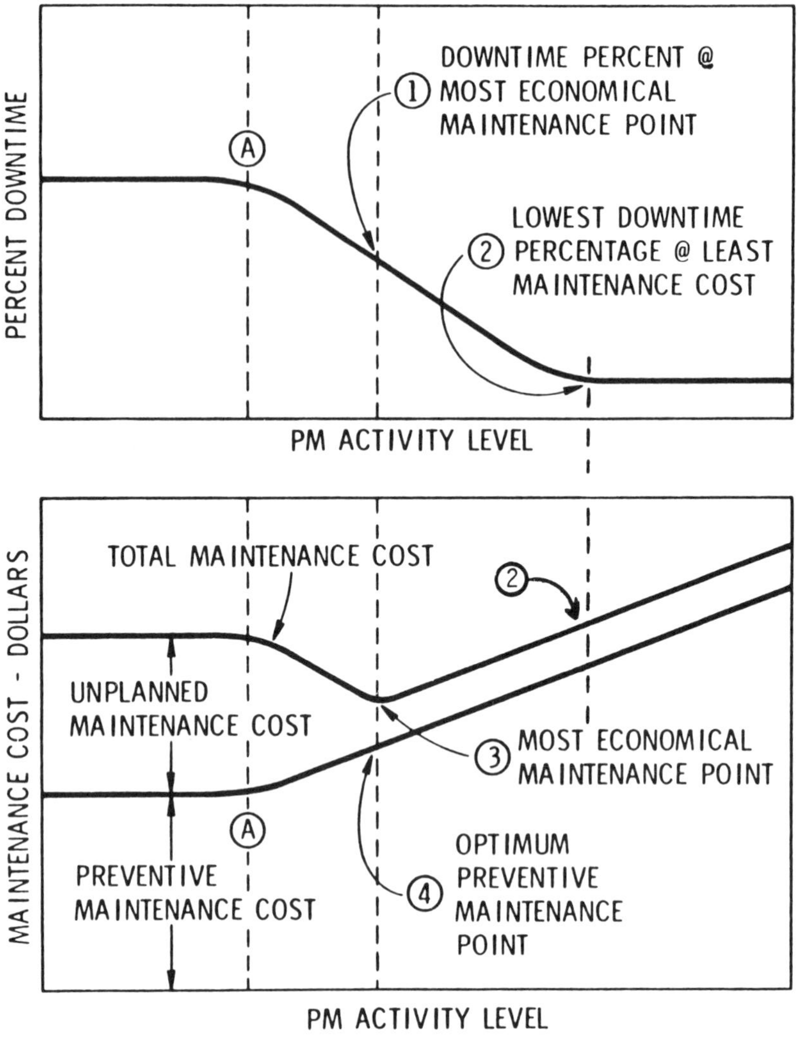

Figure 2-14: PM optimization.

How can the chart be interpreted? First, some definitions are in order. Preventive maintenance cost is the cost of maintenance resulting from preplanned, written maintenance instructions. Unplanned maintenance costs are all other related costs and are evidence of the failure of the written preventive maintenance instructions. Downtime percentage is the percentage of available machine time that the machine is broken down and unavailable for use. Construction, operation and modification costs are excluded from consideration.

Reading from left to right on all three curves up to Point A represents the established maintenance system, and includes both preventive maintenance and unplanned maintenance costs with a certain percentage of downtime. An analysis of the breakdowns in the historical record file maintained on each piece of equipment serves as a basis for improving, adding, or even reducing preventive maintenance instructions.

For example, using the automobile analogy again, if spark plugs fail frequently with consequent downtime, it would be logical to add a written preventive maintenance instruction to replace the spark plugs at periodic intervals to prevent breakdown. Point A represents the point at which additional preventive maintenance instructions are included in the maintenance program. If the original preventive maintenance program were modest in scope, and if the added instructions were intelligently written, several trends should become evident. One is that preventive maintenance costs will rise in direct proportion to the cost of implementing the additional instructions. However, the unplanned maintenance cost and the downtime percentage should drop, as well as the total maintenance cost. The reason is that planned preventive maintenance is more efficient. An individual assigned to the work is given all the instructions, materials, tools, etc., required to do it quickly. In the case of unplanned maintenance, the worker must take time to identify the problem and diagnose the fault before he can begin repairs. Frequently, he must return to the shop to get the needed material and tools which may not be in stock, further increasing downtime and costs.

Further, unplanned breakdowns often pyramid into additional breakdowns with resulting higher cost. For example, an unplanned failure of a motor bearing may cause the shaft to drop, and seizing of the rotating part. The result might be a cost rise from a $50 bearing replacement charge to a complete rebuilding cost of thousands of dollars. If preventive maintenance instructions are continually added and frequencies increased, the point of saturation is eventually reached and the preventive maintenance task is no longer economical (see Point 3). At this time a reversal in the downward trend of total maintenance cost will occur and costs will increase in direct proportion to the added preventive maintenance cost.

Point 3 is the most economical maintenance point; Point 4 is the optimum preventive maintenance point. Point 1 is the downtime percentage produced at the most economical point. Determination of the points at which economy is maximized is the purpose of the study and can result only from an intelligent analysis of maintenance records, coupled with an analytical approach to varying the preventive maintenance effort.

In some cases where critical equipment is involved, maintenance cost is of secondary importance, the lowest possible downtime being the primary consideration because it affects production loss. Point 2 represents the lowest possible total maintenance cost at which the lowest downtime can be achieved. In other words, increased preventive maintenance effort beyond this point produces no decrease in downtime. As an aid in determining the level of maintenance, it is helpful to divide facilities and equipment into three categories:

> *Category A*--Items of very small dollar value where no preventive maintenance program is planned. A $10 electric fan is an example. Here, rather than perform preventive maintenance, it is more economical to let the fan run until breakdown and then either replace or repair it. There is no preventive maintenance program for this category, only breakdown maintenance.
>
> *Category B*--Equipment of relatively large dollar value where a reasonably low downtime is required and maintenance of the equipment at the most economical point is most important. This category has a normal preventive maintenance program.
>
> *Category C*--Equipment that must be operated at the lowest achievable downtime because of the criticality of function or cost of production downtime. This category must have a special preventive maintenance program, plus an aggressive maintenance optimization program.

With this approach, system and paperwork complexity are balanced against need. Where there is no need, the system and paperwork are at zero. Where there is a requirement for normal preservation of facilities and a reasonable downtime, a preventive maintenance program is installed. Where production lost time costs are of overriding importance, a maintenance optimization system is used.

Normally, zero downtime can be achieved only by placing redundant units side by side and so interlocking them that one unit starts up if the other fails. In this case, a decision must be made whether to maintain the equipment at its most economical point or minimum downtime point.

This statement is true for equipment running continuously. However, if the equipment can be shut down for preventive maintenance when it is not being used, and if you define downtime only as time broken down during a period when the equipment is needed, then it is possible to achieve zero downtime by the intelligent use of PM. For example, if a belt normally lasts for one year, then if you renew the belt twice a year, it is unlikely that you will suffer downtime from a belt failure. Zero downtime is discussed in more detail at the end of Chapter 12.

With the maintenance cost curves analyzed, effective preventive maintenance instruction can now be prepared. If a maintenance man who is familiar with a particular machine were asked to write a set of preventive maintenance instructions which might achieve Point 3 (lowest total maintenance cost) and yet another set to achieve Point 2 (lower downtime), he probably would be unable to do it. It would take a miracle to come up on the first try with the optimum instructions to satisfy each objective. Although he might come close to it by chance, he would never know it if the cost and downtime data were not plotted on graphs such as those in Figure 2-14.

Unfortunately, this is seldom done. Little cost analysis is applied to most plant maintenance programs to determine how economically efficient the individual machine maintenance program is. The dollars wasted through lack of such analysis can add up to significant costs.

Optimization of preventive maintenance, as represented by the finding of points of economy 1-4, can be accomplished through experimentation and with the help of a well-qualified maintenance man who is familiar with the particular machine under study. It must be done in the following manner:

Starting at Point A, the historical record file should be analyzed and preventive maintenance instructions added to prevent future failures of the type noted in the record. In addition, if a preventive maintenance instruction already covers the point of failure, but breakdowns still occur, the preventive maintenance frequency should be increased.

Conversely, if a preventive maintenance instruction covers an item or function on which there has been no breakdown for a long period, either the preventive maintenance frequency should be reduced or the task eliminated. By varying preventive maintenance frequencies and tasks, and by plotting the effect on total maintenance costs, you will be able to determine Point 3. The set of preventive maintenance instructions which are in use at Point 3 will be the instructions that produce the most economical total maintenance cost.

By continuing this variation and increasing instructions and frequency past Point 3, Point 2 will be reached. This represents the lowest maintenance cost at which the lowest downtime can be accomplished. The maintenance manager is thus able to maintain his equipment at either the most economical point or the point of least downtime with corresponding minimum cost based on the criticality of the equipment.

Mandatory Vs. Non-Mandatory

Maintenance instructions can be divided into mandatory and non-mandatory types. Again, using the automobile analogy, spark plug maintenance can be considered to be an example of a non-mandatory type.

Consider the fact that an eight-cylinder car has eight spark plugs which are not likely to fail simultaneously. Actually, it is an infrequent occurrence when only one fails completely. Rather, they all begin deteriorating at the same time. If there were no specific instructions calling for periodic renewal of the spark plugs, the result would merely be a gradual decline of engine performance until it was finally found necessary to change the spark plugs. Because nothing serious will occur if you do not change the spark plugs on a prescribed or planned basis, the instruction is considered to be non-mandatory.

On the other hand, an example of a mandatory item would be the brakes on a car. If there were no preventive maintenance system on the brakes, there would be no assurance that brake failure would not occur without warning with the resultant possible loss of life. Therefore, it is mandatory that you write a planned maintenance instruction calling for periodic brake inspection and an all-out overhaul of the system after so many miles of operation. Understanding of this categorization and an intelligent appraisal of historical feedback information will help the preventive maintenance engineer make proper decisions concerning preventive maintenance coverage.

Pyramiding Vs. Non-pyramiding

Another type of maintenance instruction is known as pyramiding and non-pyramiding. Consider the following example. A fuse blowing in an electric circuit is an example of a non-pyramiding type of breakdown. When a fuse blows, that is the end of it--nothing else happens. An example of pyramiding type of failure might be the previously cited failure of a bearing--the main bearing in an engine, for instance. If it fails, the shaft may drop and the entire engine may seize, thereby causing catastrophic failure and increased cost. Understanding pyramiding will aid the preventive maintenance engineer in making decisions concerning relative importance of maintenance tasks.

Inspection Vs. Positive Instructions

There is much controversy concerning simple inspection versus "positive" maintenance instructions. Many feel that a preventive maintenance program should consist of inspection items only and very minor maintenance work. Others feel that preventive maintenance should consist only of "positive" tasks.

To illustrate the difference between the two, consider the maintenance of a belt drive. If planned maintenance were handled by inspection only, the task would call for inspection of the belt drive at a certain frequency. The instruction would probably say "Inspect the belt to see if it is in good condition and has the proper tension. If it is in bad condition, replace it or if the tension is incorrect, make an adjustment." On the specific "positive" task basis, the instruction would merely call for replacement of the belt at a specified time interval.

Again, an understanding of these two categories will enable the preventive maintenance engineer to make an intelligent appraisal of the maintenance needed. The belief that preventive maintenance must consist of either 100 percent positive tasks or 100 percent inspection tasks is unsound. Each maintenance requirement should be considered individually. In certain cases, an inspection instruction is the correct approach; in others, a positive instruction is preferable.

Returning to the belt drive example, and using the inspection approach, assume historical records show that the belts have been failing on the average of once per year. The inspection might be prescribed once per month and require round-trip travel time of 20 minutes, plus two minutes on the job. A yearly belt replacement cost of $60, in this instance, adds up to an annual total of $300 for inspection and belt replacement.

In the case of a positive instruction, belt replacement might be specified every nine months at a yearly cost of $90. In this case, it is apparent that the positive instruction is more economical. Another example of this is shown in Figure 2-8.

The following paragraphs divide the task of writing preventive maintenance instructions into its several subelements for the purpose of making clearer the various processing steps and their logical sequencing.

PREVENTIVE MAINTENANCE SUPERVISOR (PMS)

It is essential that a single person be designated to manage the preventive maintenance program. This is a full-time assignment which should be given to one of the best men in the department. The preventive maintenance supervisor is responsible for managing the implementation, operation and control of the preventive maintenance system which includes:

- o Maintenance of the criticality coding list with the help of engineering.
- o Operation of the failure recording systems.

- o Development and maintenance of the failure coding system.
- o Preparation and follow-up of critical problem reports.
- o Preparation, checking, training of personnel and scheduling of the implementation of daily inspection route sheets.
- o Preparation, checking, training of personnel and scheduling of the implementation of preventive maintenance instructions.
- o Coordination with the computer department on problems related to the computerization of the preventive maintenance system.
- o Analysis and correction of maintenance problem areas.
- o Selection, training, scheduling, expediting and providing guidance to personnel involved with the preventive maintenance system.
- o Reporting on status of the preventive maintenance systems to the maintenance manager.
- o Coordination of preplanned outages.
- o Scheduling of analysis and PM instruction writing work loads.
- o Conducting of simultaneous downtime studies.
- o Maintenance and analysis of BM/PM ratios.

IDENTIFY EQUIPMENT

The first step is to identify equipment to be included in the preventive maintenance program and the priority of program implementation.

Categorization of equipment by criticality should be accomplished as explained in Chapter 5.

The most critical equipment items should be implemented first.

The preventive maintenance supervisor should then prioritize the items that are most critical to determine which to work on first.

ESTIMATE INSTRUCTION WRITING TIME

A rough estimate of the instruction writing time should be made for each of the prioritized items by the preventive maintenance supervisor for scheduling purposes.

ASSIGN RESPONSIBILITY

At a meeting chaired by the preventive maintenance supervisor with representatives of the maintenance, production and engineering departments, specific people should be assigned to write instructions for each of the equipment items.

There are two types of instructions, the Daily Instruction Route (DIR) and the Preventive Maintenance Instruction (PMI). The same man should be assigned to write both for the same equipment item.

The writing effort should involve people of the following types:

Maintenance foreman

Maintenance staff

Engineering

SCHEDULE WRITING TASK

The preventive maintenance supervisor shall then prepare a schedule of the preventive maintenance instruction writing task utilizing the following information:

- o Priority listing of the critical items
- o Estimated hours per equipment items
- o Part-time capacity of assigned individuals

The preventive maintenance supervisor will be responsible for:

- o Maintaining a Gantt chart of progress.
- o Submitting weekly progress report to the maintenance manager.
- o Expediting task accomplishment.
- o Acting as a preventive maintenance consultant to assigned personnel.
- o Checking completed instructions.

IDENTIFY AREA RESPONSIBILITY

Maintenance responsibilities should be assigned and coded as follows:

Area Maintenance Responsibility Codes

Code	Area	Foreman
1	Maintenance Area 1	Jones
2	Maintenance Area 2	Smith
3	Maintenance Area 3	Smicek
4	Maintenance Area 4	Collins
5	Maintenance Area 5	Martello
6	Maintenance Area 6	Bauer
7	and so on	

These code numbers should be the numbers used for the first digit of the Work Order Number. The purpose of this is twofold:

- o To assign specific maintenance responsibility by area and foreman.
- o To permit computerized segregation of preventive maintenance instruction and scheduling by area and foreman.

Specific names of foremen should be assigned to each of the areas. *They* shall be responsible for the maintenance program in their assigned areas. The preventive maintenance supervisor shall be construed as being staff support to these people.

ASSEMBLE REFERENCE MATERIAL

The writer of a preventive maintenance instruction for a specific item of equipment should systematically assemble reference material as follows:

- Manufacturer's operating and maintenance manuals
- Recommended spare parts list
- Historical breakdown repair file
- Correspondence
- List and description of design changes
- Discussion notes from area maintenance supervisors
- Discussion notes from shift supervisors

The writer should study *all* of the above information items to gain familiarity with the machine and its problems.

WRITE DAILY INSPECTION ROUTE SHEETS (DIRS)

Certain preventive maintenance tasks should be performed on a daily basis. These tasks are normally of a very minor nature, many of which can be performed while the equipment is running. These tasks represent a fixed, routine and repetitive workload. For this reason, a different system is used to schedule this type of work than the system used for odd frequency preventive maintenance tasks.

Daily inspections are performed to satisfy ourselves that critical equipment is functioning satisfactorily and that situations aren't developing that might precipitate pyramiding failures and downtime. For example, if a certain pump was considered to be critical, it *might* be decided to inspect it every day as follows:

- o *Feel* all bearings for proper temperature.
- o Check that gland leakage is not excessive.
- o Check for abnormal vibration.
- o Check oil level in bearings.
- o Etc.

Each of the area maintenance supervisors should have the responsibility of developing DIRS on the critical equipment items in their respective areas. This task should proceed as follows:

- o Mark on the plot plan which of the items are in the specified area.
- o Divide the area into the logical number of routes.
- o Determine and mark on the plot plan the sequencing of inspections.

The following drawing illustrates this procedure (figures and routes are hypothetical rather than actual).

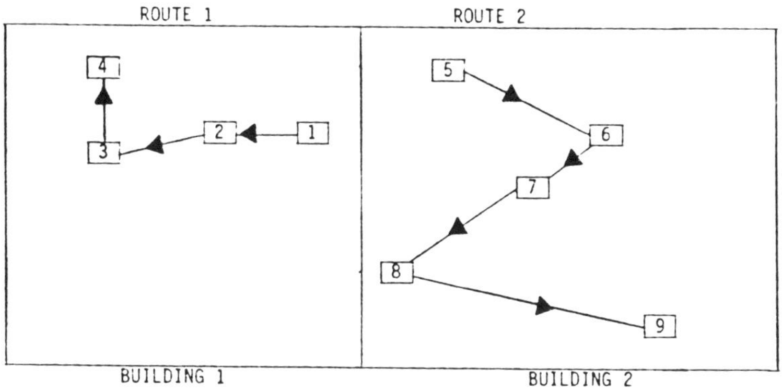

The numbered squares show that nine of the critical equipment items are located in Buildings 1 and 2.

The next step is to write inspection instructions on the form shown in Figure 2-15.

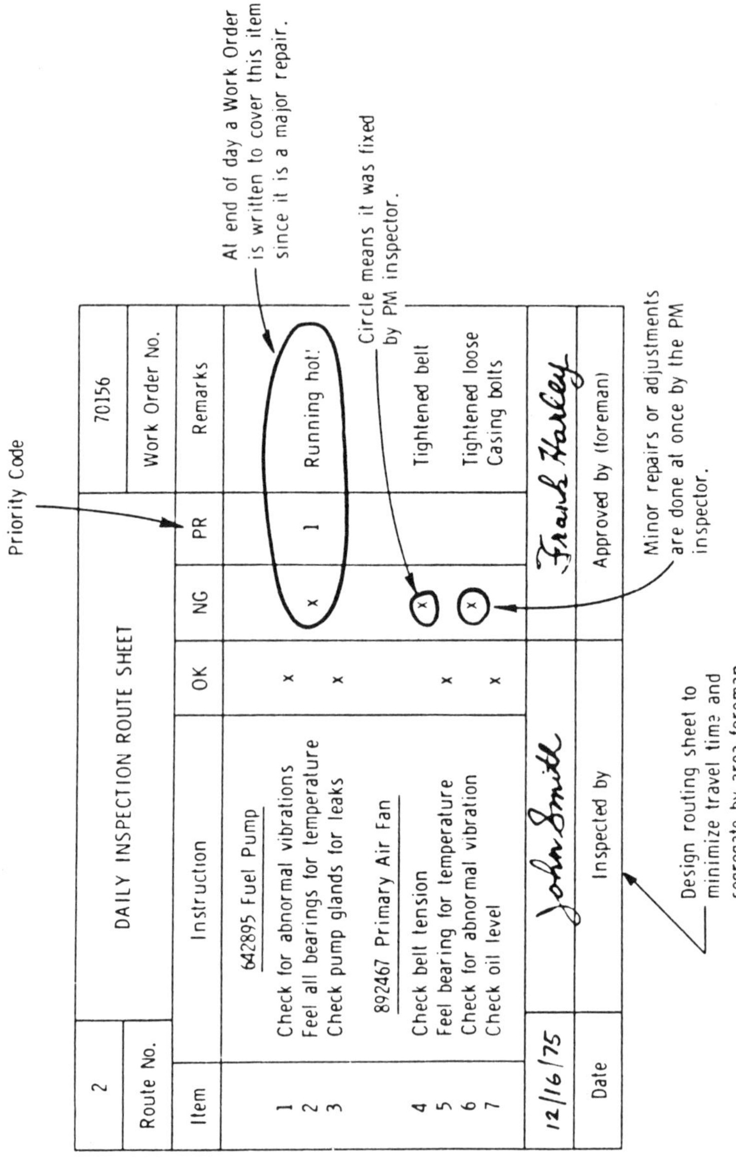

Figure 2-15: Daily inspection route sheet.

The first *Daily Inspection Route Sheets* developed should be construed as a "first-cut." In keeping with the intent of the "first-cut" effort, the route sheet writer should restrict his instructions to *obviously* essential items. Marginally effective instruction items can be given more careful thought when the writer assigned to a specific equipment item is working on the main PM instructions.

The importance of determining the optimum routing of tasks is illustrated in the following case study:

CASE STUDY 2-N: The Connect/Disconnects

While doing a study of a utility company, I was asked to look into the efficiency of the connect/disconnect operations. These tasks can be defined as follows:

> *Connects*
>
> When a new customer ordered electrical service, a man was dispatched to connect the service. This case study refers only to connections to private homes which constituted the overwhelming majority of tasks. The home would already have the service run to a meter box on the house. All the electrician would have to do is to open the box and push a connector into the receptacle. This constituted a connection.
>
> *Disconnects*
>
> A disconnect was simply the same task in reverse. The electrician was dispatched to the home and he pulled the connector out of the box.

I was therefore surprised to find out that average time taken to do disconnects was *three times more than to do connects*! This startling fact was discovered during my study. I also found out the reason at the same time. The reason was that time was taken to prepare the optimum routing for connects. However, the philosophy for disconnects was that the customer did not pay their bill and therefore they were going to dispatch a man immediately after the decision to disconnect was made without taking the time to batch and route disconnects in a manner similar to connects. They were biting their nose to spite their face! I proved to them that any savings in cost of electricity was dwarfed by the inefficiency of operations.

This case study dramatically illustrates the need to determine the optimum routing of DIRS in industrial plants as well as utilities.

WHO SHOULD PERFORM DAILY INSTRUCTION SHEET TASKS?

The following personnel are candidates for performing daily inspection route sheet tasks:

- Production supervisors
- Maintenance foremen
- Production workers
- Maintenance workers

Depending on the circumstances, one or the other of these groupings may be the more logical choice. For example, *if* a production worker is specifically assigned to watch a piece of equipment, and *if* the daily inspection sheet is within his technical capability, and *if* the task does not interfere with his normal assignment, then *he* may be the most logical choice.

It is up to the instruction writer to make recommendations concerning this matter. There may be cases where the most economical choice is to break up route sheet tasks to several of the four groupings.

PERFORMING DAILY INSPECTION ROUTE SHEET TASKS

The person making the inspection performs the work in the order shown on the instruction form.

After inspecting the item he puts an "X" either in the OK or NG (no good) column, as appropriate.

If the item is "NG" and the fix required is minor and within the skill limitations of the inspector, he shall fix the item on the spot. By "minor," we are referring to minor adjustments, such as, tightening nuts or bolts, tightening gland seals, adding oil, adjusting belts or other items, etc. It is not intended that he spend more than 2 hours per day on "fixing" things.

Whenever an item is marked "NG," an explanation is given in the remarks column to clarify the problem. If a "fix" is made, a circle is drawn around the "X" in the "NG" column signifying that problem was corrected on the spot. Discrepancies that are beyond the inspector's capability to fix are left uncircled. In this case, the inspector indicates his appraisal of the seriousness of the problem by inserting a priority code number.

Examples of the various cases are shown in Figure 2-15.

At the end of the day, the inspector dates and signs the sheet indicating completion of the task. This is a very important step. Inspectors should be made to understand that signing off as complete, items that were not performed, is a cause for dismissal. Performance against these tasks should be periodically audited by the area foreman.

If a discrepancy is considered to be an emergency situation and the inspector cannot fix it himself on the spot, then he should notify the area foreman or the maintenance office immediately.

After the foreman receives and notes discrepant items on the route sheet, the sheet is turned over to the chief planner scheduler who shall:

- o Write special work orders for the uncircled discrepant items.
- o Plan and schedule these work orders.
- o File route sheets chronologically by area and route number. These shall be held for a year for auditing and analyses purposes, then discarded. The preventive maintenance supervisor should review the check marks during the annual audit. Instructions with *zero* "NG" check marks should be considered

for cancellation. Instructions with *very* frequent "NG" checks should be candidates for revision.

CHARGING ROUTE SHEET WORK

A blanket work order should be written for each daily inspection route. All work in this category should be charged to the appropriate blanket work order.

PREVENTIVE MAINTENANCE INSTRUCTIONS (PMI'S)--GENERAL

All frequencies above daily should be specified and scheduled on preventive maintenance instructions. This work will constitute the major portion of the preventive maintenance system workload.

Since there will be many different equipment items, frequencies and instructions, the scheduling task can be complex and confusing unless a special scheduling system is designed. Since most companies now have computers, the system described in the following paragraphs is a computerized system which utilizes a simple sorting program and is designed purposely to simplify the scheduling function.

The writing of preventive maintenance instructions for the critical equipment items starts with an appraisal of the existing situation. If there are no breakdowns or an insignificant number, concurrently with low or reasonable maintenance costs, there may be no need to write preventive maintenance instructions.

However, if breakdowns are high, concurrently with excessive maintenance and production lost time costs, preventive maintenance instructions are probably justified.

The first question that a preventive maintenance instruction writer should ask is "What am I trying to achieve?"

The answer to this question can be found in Figures 1-2, 2-3 and 2-14. We are trying to achieve on each piece of equipment one or more of the following:

- o Lowest combined cost of maintenance and production loss caused by downtime
- o Lowest or zero downtime
- o Most economical PM cost

The next question is, "Suppose there is never or seldom a production loss when an equipment item breaks down, what then?" In this case, we are looking for the least combined cost of breakdown and preventive maintenance.

The writing of intelligent preventive maintenance instructions is the most important and difficult assignment in the preventive maintenance program.

The most common fault is to underestimate the skill required to write preventive maintenance instructions. Preventive maintenance instructions improperly written can scuttle a preventive maintenance program by increasing costs, and manpower requirements without off-setting decreases in production lost time costs. The knowledge required to properly write preventive maintenance instructions is acquired by a lifetime of exposure to maintenance problems and their solutions. The preventive maintenance instructions should give full consideration to the philosophies espoused in this chapter concerning positive versus inspection, mandatory versus non-mandatory, pyramiding versus non-pyramiding and Figure 2-14, when writing preventive maintenance instructions.

Furthermore, the ratio of overhaul cost to production loss cost should be considered. Large ratios should discourage major overhauls. Small ratios should encourage overhauls.

When this ratio is large, the emphasis should be on examining historical failure records and writing preventive maintenance instructions slanted towards specific highly repetitive failures rather than major overhauls.

One must also realize that preventive maintenance outage costs, potential production loss costs from lack or addition of preventive maintenance work and the specific content of the preventive maintenance instruction are inextricably interwoven. Instructions which ignore this relationship are doomed to failure.

As the preventive maintenance instruction writer writes *each* instruction, he must ask the question, "Is this instruction going to reduce the combined cost of maintenance and production loss?" If the answer is *no*, then the instruction should not be written, unless safety is a factor.

In most cases, it will be fairly obvious whether or not a preventive maintenance instruction is worthwhile. In some cases, it may not be. As a matter of fact, the answer of whether or not it will pay off may be indeterminate. In this case, the preventive maintenance instruction writer must make an educated guess. The failure analysis chart, Figure 2-3, will give the final answer of whether or not the guess was correct or almost correct.

This word of caution concerning cost impact is given because when a preventive maintenance instruction is implemented, there is an immediate cost increase. We still have the existing costs until the preventive maintenance instruction has a chance to exert its cost reducing effect.

For this reason the preventive maintenance instruction program should be implemented gradually. When the preventive maintenance instruction on the first of the critical machines is written, it should be immediately implemented. The others should be implemented as they are written.

When the preventive maintenance instructions on all critical machines are implemented, there should be a pause for at least one month, during which time the preventive maintenance supervisor should assess the improvement process. When he is satisfied that there has been an actual cost reduction, he can start implementing machines in a lessor criticality category.

In addition, the preventive maintenance instruction writer should have a thorough understanding of the principles of failure analysis given in Chapter 10. These principles affect not only the writing of preventive maintenance instructions but also the revisions, updating and improvements to preventive maintenance instruction which are a natural part of the maintenance analysis process.

PREVENTIVE MAINTENANCE INSTRUCTION COMPUTER INPUT FORM

The form to be used in writing preventive maintenance instructions is shown in Figure 2-16.

Sheet 1 of 1

682958	P. M. INSTRUCTION	3
MACHINE NUMBER		FOREMAN
MACHINE DESCRIPTION		COST CENTER NO.

ITEM	INSTRUCTION	MECHANICAL					ELECTRICAL					LUBRICATION				
		W	M	Q	S	A	W	M	Q	S	A	W	M	Q	S	A
1	Inspect Oil Level, add Mobil XYZ, if necessary.	.2														
2	Renew Inlet Filters.			.2												
3	Open Fan Inspection Plate; inspect and report findings.					.3										
4	Renew Belts.					.3										
	SCHEDULING CODE NO.															
	TOTAL HOURS	.2		.2		.6										

Figure 2-16: PMI computer input form.

Notice that the form is segregated for three trades:

- o Mechanical
- o Electrical
- o Lubrication

Other types can be added, if necessary.

Each of the three trades have columns for the various frequencies which are:

W - Weekly

M - Monthly

Q - Quarterly

S - Semi-annually

A - Annually

A rough estimate should be entered as shown opposite each instruction item. The purpose of this estimate is to give an approximate idea of workload for scheduling purposes. The estimate should be given in hours and tenths of an hour.

A blanket work order number should be assigned to each preventive maintenance instruction. *All* work performed against the preventive maintenance instruction should be charged to the blanket work order number. (The machine number put on the blanket order will be the specific machine for which the preventive maintenance instruction was written.)

LOAD BALANCING

It is advisable to spread the preventive maintenance instruction workload as evenly as possible throughout the year in order to avoid disruption of other maintenance work. The scheduling code number is a computer tool for accomplishing this. A space is provided on the preventive maintenance instruction sheet for this code. See Figure 2-16.

There are several considerations affecting the date chosen to perform the preventive maintenance instruction, i.e.:

- Seasonal shutdown of parts or all of the plant.
- Need to leave one redundant unit operating while the other is given its preventive maintenance instruction.

- o The logic of taking certain groupings of equipment out of service at the same time.
- o Other reasons of a similar nature.

There are common sense considerations that should be made by the PM instruction writer and tailored to each individual plant.

The load balancing function is facilitated by use of a table as shown in Figure 2-17. A separate table is made for each trade. This table is filled in by hand. The steps in using the table and schedule code number are:

Week → / Equip. No. ↓	1	2	3	4 through 51 →	52
682958	.2 .2	.2	.2		.2 .6
TOTALS	.4	.2	.2		.8

Figure 2-17: PMI load balancing chart (mechanical).

Step 1: Enter the hours and equipment number from the total column of the preventive maintenance instruction for the first machine. Notice that the table has the 52 weeks in the year in vertical columns and equipment numbers in horizontal columns. The total hours are continuously updated as each new machine is added to the table.

Referring to the preventive maintenance instruction Figure 2-16, we notice that there is a weekly task of .2 hour. Therefore, we enter .2 hour for each of the 52 weeks. Then there is a quarterly task of .2 hour. Now we have a choice of when we perform this task. Let us choose the 1st, 14th, 27th and 40th weeks. The .2 hour should be entered into the chart opposite the appropriate week as shown. Finally, let us choose the 52nd week to perform the annual task and enter the amount as shown. In the case of the annual task, we have 52 choices.

Step 2: Select the scheduling code from Figure 2-18. We can see from this figure that there are the following choices of codes depending on the task frequencies.

Frequency	Number of Choices
Weekly	1
Monthly	4
Quarterly	13
Semi-annually	26
Annually	52

The code number is arranged as follows:

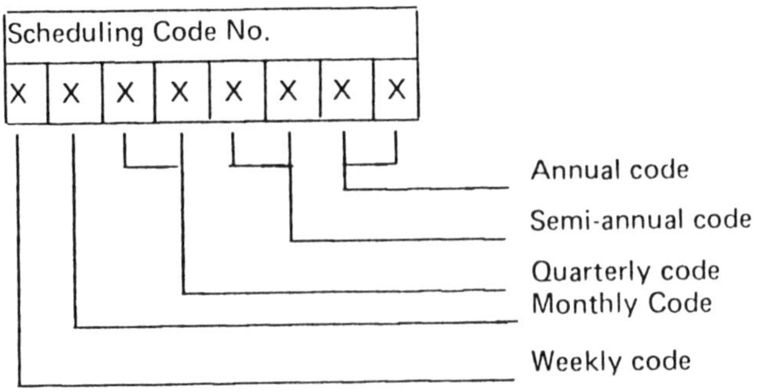

WEEK	SCHEDULING CODE				
	WEEKLY	MONTHLY	QUARTERLY	SEMI-ANNUALLY	ANNUALLY
1	1	2	06	19	45
2	1	3	07	20	46
3	1	4	08	21	47
4	1	5	09	22	48
5	1	2	10	23	49
6	1	3	11	24	50
7	1	4	12	25	51
8	1	5	13	26	52
9	1	2	14	27	53
10	1	3	15	28	54
11	1	4	16	29	55
12	1	5	17	30	56
13	1	2	18	31	57
14	1	3	06	32	58
15	1	4	07	33	59
16	1	5	08	34	60
17	1	2	09	35	61
18	1	3	10	36	62
19	1	4	11	37	63
20	1	5	12	38	64
21	1	2	13	39	65
22	1	3	14	40	66
23	1	4	15	41	67
24	1	5	16	42	68
25	1	2	17	43	69
26	1	3	18	44	70
27	1	4	06	19	71
28	1	5	07	20	72
29	1	2	08	21	73
30	1	3	09	22	74
31	1	4	10	23	75
32	1	5	11	24	76
33	1	2	12	25	77
34	1	3	13	26	78
35	1	4	14	27	79
36	1	5	15	28	80
37	1	2	16	29	81
38	1	3	17	30	82
39	1	4	18	31	83
40	1	5	06	32	84
41	1	2	07	33	85
42	1	3	08	34	86
43	1	4	09	35	87
44	1	5	10	36	88
45	1	2	11	37	89
46	1	3	12	38	90
47	1	4	13	39	91
48	1	5	14	40	92
49	1	2	15	41	93
50	1	3	16	42	94
51	1	4	17	43	95
52	1	5	18	44	96

Figure 2-18: Scheduling code.

Since we have already determined which week each task is to be performed, we can then get the code number from Figure 2-18 which is:

1 - 0 - 06 - 00 - 52

The zeros in the monthly and semi-annual spaces indicate that there are no instructions for these frequencies for the first machine.

Step 3: Each week the computer will observe the scheduling code numbers "valid" for that week, since it has a copy of Figure 2-18 stored in its memory.

For example, for the 1st week of the year the "valid" code numbers are:

1 - 2 - 06 - 19 - 45

The computer then compares the scheduling code for the specific machine and determines where there is a match. Using the preventive maintenance instruction code previously determined and the 1st week, we have:

Valid numbers	1	-	2	-	06	-	19	-	45
PMI code	1	-	0	-	06	-	00	-	52
Match ? Yes or No	Yes		No		Yes		No		No

In this case there is a match for the weekly and quarterly instruction, therefore, the computer will print out instructions to do these tasks as shown in Figure 2-16.

Step 4: When the second item of equipment is ready to be coded, we consider the load spreading features of the scheduling code and its choices. Now when we select the various numbers to make up the schedule code, we will select codes to spread the load as evenly as possible throughout the years--52 weeks. This is advisable since we would like to have a fixed size crew to handle preventive maintenance to the extent possible. More will be said about scheduling preventive maintenance in Chapter 16.

SCHEDULING

Each week on Friday afternoon, the computer shall print a set of instruction cards which specify the preventive maintenance tasks that must be done the following week. Figure 2-19 shows the form to be used by the computer. The cards should be printed in duplicate and segregated by:

- o Foreman
- o Mechanical Work
- o Electrical Work
- o Lubrication Work

In other words, no single card should contain instructions for more than one trade category or foreman.

The two sets of cards properly segregated should be distributed to each area foreman, so that he can plan his preventive maintenance work for the next week.

The foreman should retain one set of cards for a follow-up. The other set should be given to the workers so they can perform the work.

When the worker finishes the task specified on the card, he should initial the box labeled "complete" signifying that he has completed all tasks on the cards.

If he completes each task exactly as specified, he puts a check mark opposite the item. If unanticipated trouble was found or if he wishes to comment on an item, he puts an "X" mark opposite the item and puts his comments on the reverse side with a appropriate item number reference. For example, in the "X" marked item in Figure 2-19, the comment might read:

> 3--Fan impeller badly corroded. Should be replaced

When the hourly worker returns the signed-off card to the foreman, he matches it with his copy; if the card contains all check marks, he throws away both copies (after noting time card charges).

P M I	682958	Primary Air Fan	F.W. [illegible]	M	26	2	3
	Mach. No.	Description	Complete	Trade	Week	Foreman	

No.	F	Instructions	Hr	√ x
1	W	Inspect oil level; add mobil XYZ, if necessary	.2	√
2	O	Renew inlet filters	.2	√
3	A	Open Fan inspection plate insp.; report findings	.3	x
4	A	Renew belts	.3	√
		see remarks for items marked "x" on reverse side		
		Total		

(BACK)

3. Fan impeller badly corroded. Should be replaced.

Figure 2-19: PMI scheduling card.

If the mechanic's card contains an "X," the foreman throws away his copy, notes the discrepancy and forwards the mechanic's card to the chief planner scheduler for action.

Any cards left over with the foreman indicates that a card given to a mechanic has been lost or destroyed. The foreman must take appropriate action in the case.

More details on the scheduling of preventive maintenance instruction *after* the cards are received by the foreman are given in Chapters 15 and 16, Planning and Scheduling.

COMPUTER FUNCTION--REPORTING

In summary, the computer will record and print out costs for preventive maintenance work in the same manner as other work, by means of time cards and work orders. The following table summarizes the various cases of preventive maintenance charges:

Type of PM Order	Remarks
PMI work	Blanket work orders charged to specific equipment
Daily inspection route sheets	Blanket work orders charged against route numbers
Discrepancies surfaced by PMI's or DIRS's	Specific work orders charged against specific equipment

PREVENTIVE MAINTENANCE WORK PERFORMANCE

Concerning the performance of preventive maintenance work, the most important thing to remember is that the mechanic is in the most advantageous position to judge the quality of preventive maintenance instructions. He should not only perform each preventive maintenance task as directed, but should also be *specifically* instructed to maintain an open mind in judging the effectiveness of each preventive maintenance instruction. Some of the items for him to look for are:

o Do frequencies appear to be too high, too low, or just right?

- o Are repetitive breakdowns occurring on parts or systems *not* covered by preventive maintenance?
- o Should certain inspection tasks be changed to positive tasks or vice-versa?
- o Should certain tasks be eliminated?

Foremen should periodically cover these points with the mechanic in a mutual discussion designed to draw out ideas for improvement from the mechanic.

AUDITING OF PREVENTIVE MAINTENANCE WORK

Foremen should periodically audit the mechanic's performance of preventive maintenance tasks to ensure that all tasks are completed as signed off.

Signing off on an incomplete task is a serious *offense.* This point should be made quite clear.

Foremen should remind mechanics that they are expected to make minor adjustments and repairs while in the process of performing work.

The audit should be performed as frequently as is necessary for the foreman to have confidence that the mechanic is not only performing each preventive maintenance task scheduled, but that he is actively engaged in analyzing preventive maintenance tasks and discrepancies with a goal of constant improvement to written preventive maintenance instructions.

The mechanic should be reminded that the PMI task is critical to the company's financial success and therefore to the security of his position.

Part 3

Productivity Control

14

Factors Affecting Productivity

GENERAL

Some of the factors affecting productivity are:

- o Crew size and mix
- o Material availability
- o Tools and equipment availability
- o Travel time
- o Job planning
- o Outage planning
- o Working conditions
- o Job method
- o Employee morale
- o Scheduling
- o Employee skills

Each of these factors must be controlled basically as follows:

- o Establish policy.
- o Measure performance against policy.
- o Take corrective action when deviation between policy and performance occurs.

The manner in which the control is used will vary for each of the eleven factors affecting productivity, but the principle will be the same. An example of this control cycle for the productivity factor of "crew size" might be as follows:

Policy: Packing of pumps under 50 HP shall be performed by one man.

Measurement: On work order X424B, 3 men were used to pack a 10 HP pump.

Corrective Action: Foreman was advised of his error and requested to comply with policy.

In the beginning there may be no pre-set policy on certain productivity factors, however, in this case the factor should still be periodically audited on a spot sampling basis, and policy set after the fact. An example of this might be a measurement of "working conditions." An audit reveals complaints from the men and the union about the shop area which was very dirty and covered with oil, chips, dust and trash. This resulted in a policy being established to have machine operators sweep their waste material into the aisle and to have janitors collect the aisle dirt. Work sampling should be done at least annually in each maintenance area. Past work sampling studies should be used for comparison purposes after the first audit to determine if improvement has been made in the productivity factor. Suggestions for controlling each of the eleven productivity factors are contained in the following paragraphs.

CREW SIZE AND MIX

The crew size and mix of labor on a job is one of *the major* factors influencing productivity.

An example will illustrate this point. Changing the gland packing on a small pump is a one man job for the simple reason that there is only room for one man to work. If you put three men on the job, the productivity will be 33 1/3% maximum. It's as simple as that.

Foremen should be constantly conscious of the crew size problem. They should make certain that excess personnel are used on backlog jobs rather than for "loading" existing jobs beyond their economical crew size.

Policy should particularly be established on highly repetitive jobs.

Auditing of this problem is simply a matter of spot checking jobs in the field. The correct crew size is normally quite obvious.

MATERIALS--GENERAL

The availability of material required to do each job is a *major* factor affecting productivity. More specifically, it's the *time* required to obtain material that principally affects labor productivity. This time is non-productive. There are four systems that are employed in cutting down on maintenance material acquisition time:

- o Effective stocking of parts
- o Bench stock
- o Kitting
- o Effective procurement

Each foreman has the responsibility of adding new items to the stockroom as required by virtue of changing conditions. The stockroom system should automatically delete items that are no longer being used.

MATERIALS--BENCH STOCK

The free issue bench stock principle recognizes that low cost, fast moving stock items frequently account for a major percentage of the number of items used but only a minor percentage of the total inventory dollar value. For example, in one plant, 80% of the line items account for only 4% of the dollar value of stock items issued and/or stocked. For these type items the cost of processing issue slips may be very uneconomical. Recognition of this fact has led many to the bench stock principle, which provides for these materials to be placed in open bins available to the mechanic in a place convenient to their use point. This saves processing cost and mechanics time.

The following procedure can be used to determine the most "economical" number of bench stock items. The most "economical" refers to the lowest cost of the sum of inventory and material acquisition time costs.

- o Provide a special computer report (one time) that shows a year's history on *maintenance* stock items issued. The tabulation should be in the format shown in Figure 1-4 with stock items and cost substituted for equipment numbers and cost.

  This tabulation will permit curves to be drawn similar to those shown in Figure 3-1.

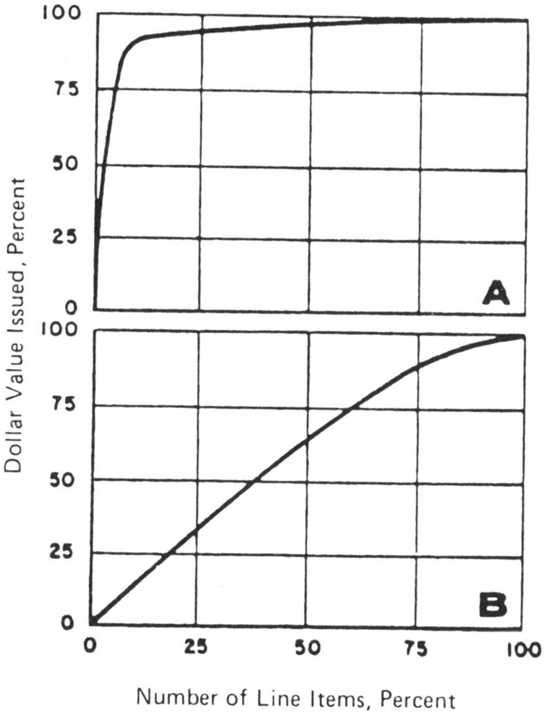

Figure 3-1: Bench stock indicators

The shape of the curve will indicate the extent to which the free issue stock principle may be employed. A curve similar to the one shown in curve A would encourage a large number of free issue items whereas curve B would discourage free issue items.

- o Convene a meeting of all maintenance foremen. In this session prepare, with everyone's help, a list of highly repetitive maintenance jobs. Divide the number of jobs on this list evenly by foremen and have them prepare a list of highly repetitive stock items used on these jobs.

- o Those items which represent fast moving low dollar value parts observed on the tab run (representing a large number of items but a very small number of dollars) and which also show up on the list of highly repetitive items used for highly repetitive jobs should be made bench stock items.

- o When contemplating where to locate the bench stock, the following alternatives should be considered depending upon the circumstances:

  - At the equipment: When the item is used for specific equipment.

  - At the center-of-gravity-of-use point: When the item is used throughout the plant. This could include one or more locations.

  - In the storeroom: In a non-secured area.

- o In addition, a decision will have to be made concerning who will replenish the bins--stores or maintenance personnel. The location and replenishing decision must be decided upon the basis of an economic analysis.

MATERIALS--KITTING

Kitting is defined as the process of obtaining all materials for a specific job in advance of the job start date. A fully kitted work order enables the job to proceed without interruption due to material unavailability. Kitting therefore increases productivity.

Construction, modification, rearrangement, and major maintenance work can be conveniently kitted due to the fact that this type of work can normally be planned well in advance of its need date. Because of this and the beneficial effect kitting has on labor productivity, this type of work should be kitted.

Referring to the list of maintenance priority codes in Chapter 9, normally Priority Codes 3 and 4 should be kitted. Priority Codes 1 and 2 (Emergency and Urgent Work) will normally not be kitted due to the lack of time. Priority Code 5 work will normally not be kitted because of the marginal justification implied in this code. This work may, however, be kitted upon approval of the maintenance manager.

Kitting should be accomplished by utilizing the following procedure:

Maintenance Responsibility

- After the work order has been prepared and approved, the planner for the particular job should prepare a bill of materials similar to that shown in Figure 3-2.
- Prepare purchase requisition for all outside purchases. (A note shall be placed on the purchase requisition to have the purchased item tagged for the work order.)
- Send bill of materials to warehouse.

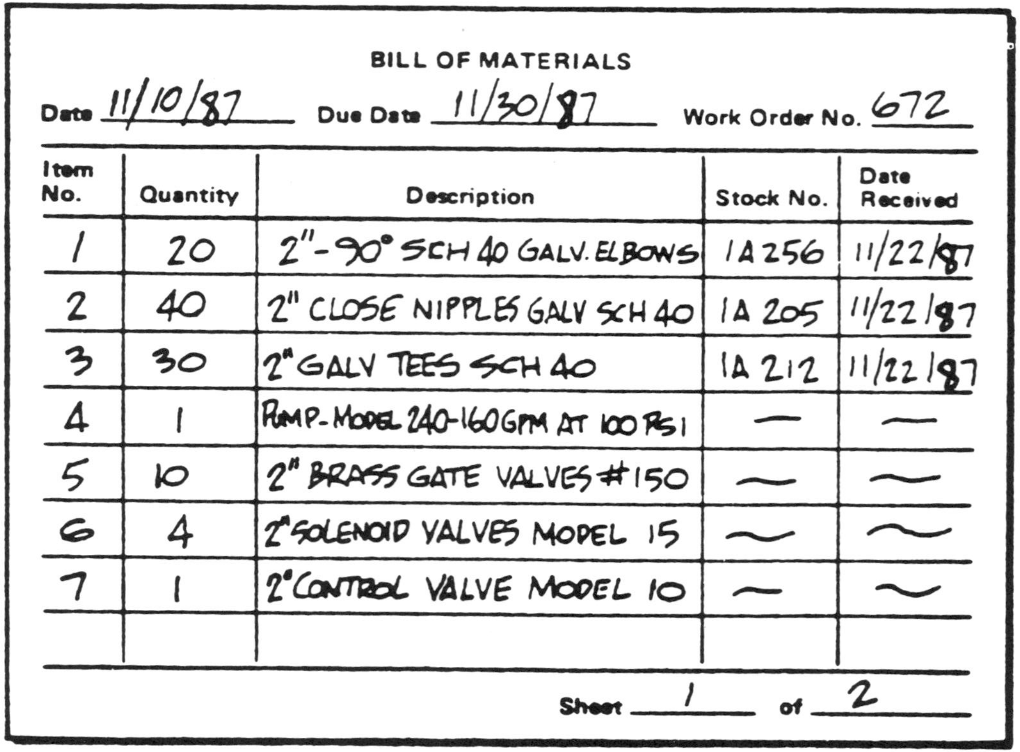

BILL OF MATERIALS

Date 11/10/87 Due Date 11/30/87 Work Order No. 672

Item No.	Quantity	Description	Stock No.	Date Received
1	20	2"-90° SCH 40 GALV. ELBOWS	1A 256	11/22/87
2	40	2" CLOSE NIPPLES GALV SCH 40	1A 205	11/22/87
3	30	2" GALV TEES SCH 40	1A 212	11/22/87
4	1	PUMP-MODEL 240-160GPM AT 100 PSI	—	—
5	10	2" BRASS GATE VALVES #150	—	—
6	4	2" SOLENOID VALVES MODEL 15	—	—
7	1	2" CONTROL VALVE MODEL 10	—	—

Sheet 1 of 2

Figure 3-2: Bill of materials.

Warehouse Responsibility

- Establish manila file folder for each work order which is to be kitted.
- Insert copies of bill of materials, purchase requisitions and purchase orders in folder.
- Write stock requisitions and fill all stock items on bill of materials.
- Designate kitting area for work order and mark location on work order kitting folder.
- Deliver stock items to designated kitting area.
- Receive, check off, and deliver outside purchases to kitting area.
- When entire bill of materials is checked off as complete, call maintenance and notify them to pick up completed kit.

MATERIALS--ACQUISITION TIME

Because material acquisition time is a major item of non-productive maintenance labor, it is worthwhile to conduct a periodic audit of this time to have some feel for the success or failure of improvements and to determine when the time spent is no longer of concern. The following procedure is recommended.

- That the maintenance planner conduct a semi-annual audit of this problem area
- Each audit shall last one week.

During each 7 day audit period, all members of the maintenance staff shall record, on a daily basis, all of the time they spend obtaining materials. "All members" include:

- *All* supervision
- *All* staff
- *All* craftsmen

It excludes:

- o Engineers
- o Operating personnel
- o Warehouse personnel
- o Purchasing personnel

All time spent in obtaining materials includes:

- o Time waiting for materials
- o Time traveling for materials
- o Time transporting materials
- o Time identifying materials
- o Time substituting materials
- o Time locating materials
- o Purchase Requisition preparation time
- o Approval time
- o Non-productive time caused by being held up by other labor without proper materials; wrong materials delivered; wrong materials ordered; material stock depleted

Time should be recorded in minutes.

The maintenance planner should prepare an audit report showing:

- o Total hours and minutes worked that week
- o Total hours and minutes spent in obtaining materials
- o Percent time spent in obtaining materials

This report will indicate the seriousness of the problem. Future reports will measure the success or failure of corrective action.

A separate but equally important consideration is the

acquisition time of materials in the purchasing department. This is discussed in Chapter 19.

TOOLS AND EQUIPMENT AVAILABILITY

We must be concerned about having proper tools and equipment on all jobs but most particularly in the following cases:

- o Highly repetitive failures
- o Failures on critical production equipment

In these cases, the highly repetitive failures surfaced by the Failure Analysis Report, Figure 2-3, should be analyzed carefully to see if any special tools, equipment, jigs or handling devices can be designed or purchased in order to reduce the repair time.

TRAVEL TIME

Travel time should be audited and controlled in a manner similar to that described for material time, if it is considered to be a problem.

CASE STUDY 3A: The Drag Line

One day I decided to see how the drag line operation was doing. We had a man working a drag line doing "ditching and diking" work around the perimeter of our plant to form the main basis for our mosquito control program.

The man operating the drag line started work at 8 A.M. and quit at 4:30 P.M. I arrived at the site at 9 A.M. and found no one. I called into central control and found out that he was on the way. He arrived at 9:45 A.M., almost 2 hours late.

He explained his lateness as follows:

"I arrive at work on time each day at 8 A.M. In accordance with my instructions, I'm to report at the central shop and wait for the bus to take me to my work station. The bus leaves each morning at 8:30 A.M. and since I'm the furthest away (20 miles), I get dropped off last. I usually arrive at my work station anywhere between 9:30 and 10 A.M. Since I'm the furthest away, I am picked up first at 2 P.M. so that I can get back with sufficient time to clean up."

It can be seen from this schedule that the man was only working 4 hours per day. (This case might be out of the ordinary since it was at Kennedy Space Center which had a huge land area.)

I immediately changed his routine to have him drive directly to the job site in his own car with an agreement to pay him mileage for the extra miles driven in his car.

Problems associated with travel time are particularly acute in very large plants that are scattered over large land areas. However, travel time can be a problem in smaller plants particularly when the company's systems and procedures force back-tracking and waste motion.

JOB PLANNING

In order for a maintenance crew to be able to proceed with a job, they must have "instructions" or a "work plan." If the work isn't planned before the crew is sent out to do a job, it must be planned during the job. There are several things wrong with the latter course of action:

- Preplanned jobs are normally planned by one person. Jobs planned during the job usually tie up the entire crew plus the foreman in the planning function. It would not be unlikely to have a planner take 1/2 hour to plan a job as compared to an 8 man crew take all day or 64 man hours to plan a job they were sent out to do without preplanning.
- There is much more likelihood to run into a stumbling block like inability to get certain parts, or an outage, etc.,

with the result that considerable idle time may be suffered. As a matter of fact, the idle time may amount to 100%, since the entire crew frequently is sent to another job because of the lack of material.

- o Planning done on a "surprise" basis usually doesn't match the thoroughness and accuracy of planning done by an experienced planner before the job is started.
- o Frequently the crew is not as competent as the planner to plan the job with the result that the planning takes longer and may not be of as high quality. Poor quality planning will further decrease productivity.

CASE STUDY 3B: The Missing Bracket

While working in an aluminum extrusion plant, a system for automating a critical machine was designed. I was excited about the design and could hardly wait to see it being installed and working, because it promised to achieve large manpower savings.

On the day installation was supposed to start, I was out at the machine at 8 A.M. I was disappointed to see the crew of 6 mechanics and one electrician arrive one hour late at 9 A.M. and made a mental note to talk to their foreman about this waste of 7 manhours.

The crew gathered around the prepared drawings. They agreed that the starting point of the job had to be the installation of an electric solenoid operated multi-ported hydraulic control valve that was the heart of the system. After this was installed then the mechanics could start running the many hydraulic lines and the electrician the wiring. "Where is the bracket to support the control valve?" asked the lead mechanic. Another said, "There isn't one shown on the drawings."

Startled, I myself looked at the drawings and apologized

for forgetting to design a bracket so that it would have been kitted and ready before the crew came out to the job. The lead mechanic told one of this men to get a welding machine, cutting torch and a length of 2" x 2" x 1/4" angle iron so that a bracket could be made. The men sat around and gossiped while waiting for the mechanic to come back with the welding machine and angle iron.

Since the mechanic wasn't back by 11:30, they all left for lunch. At 12:30 they were all back with the welding machine and angle iron. It only took 1 hour to make the bracket and the crew was now back at the starting point.

It would have taken me only 10 minutes to sketch a bracket on the drawings.

As a result of this omission, the crew was idle for 3 1/2 hours for a total lost time of 3.5 x 7 = 24.5 hours. I learned a lesson from this minor incident that I shall never forget! DON'T SEND CREWS OUT ON A JOB WITHOUT COMPLETE PLANNING IN ADVANCE!

OUTAGE PLANNING

The manner in which equipment outages are handled affects labor productivity.

Intelligent handling of this problem is complex. Suggestions for the proper handling of outages is given in Chapter 10.

WORKING CONDITIONS

Working conditions affect not only productivity but also:

- Morale
- Production loss
- Health
- Safety

Workmen should be instructed to clean up after each

job as part of their contribution to keeping the plant clean.

Foremen should see to it that their shops are not only efficient, but also have the visual appearance of being efficient.

This is important from a morale standpoint, and to promote a feeling of confidence on the part of customers of the shops.

JOB METHOD

The method used in doing a job influences productivity. The priority given to a job method improvement study on a specific job should be proportional to:

- The productivity percentage of the job
- The potential for improvement

For illustration purposes let us assume that a work sampling study has been made on a group of maintenance workers doing a specific job and the results are as follows:

	Percent
Travel Time	15
Material, tool and equipment acquisition time	50
Idle time	10
Obtaining instructions	5
Working	20
Total	100

If in this case a quick investigation reveals a reasonable job method, quite obviously the improvement study should concentrate on the excessive Material, Tool & Equipment Acquisition Time rather than on the job method.

Conversely, if a time study on a repair job showed:

	Percent
Travel	14
Material	14
Idle	10
Instructions	2
Working	60
Total	100

then a study of the job method might be warranted.

EMPLOYEE MORALE

Each supervisor has the responsibility of being concerned about the morale of his employees and for doing what he can to keep this factor high.

The art of doing this is the art of "motivation." It is the responsibility of the maintenance manager to direct a motivational program under the guidance of top management.

The end result of the program is to develop in the employee a feeling that the company:

- Recognizes him as an individual.
- Recognizes him for his accomplishments.
- Recognizes him as an important factor in the success of the company.

Great sophistication in motivational program development is not needed. A simple and very effective motivational device, for example, is to slap an employee on the back periodically and thank him for a job well done. The three types of recognition listed above must be communicated and communicated repeatedly. These motivating communications are like a food that a person must have to survive and be productive.

A company or supervisor who does nothing positive in the art of motivation must inevitably suffer low morale and low productivity.

CASE STUDY 3C: "A Day's Pay for a Day's Work"

At one plant, it soon became obvious that the morale of practically all maintenance workers was very low. Their productivity as determined by work sampling and spot audits was correspondingly low. After being there for a month I found many reasons for the low morale.

What was most shocking to me was that I could find no one responsible for, or aware of, a motivation program. I was particularly surprised to find that the director of personnel said he was not responsible for a motivational program nor did he know of anyone who was. (This responsibility is normally given to the director of personnel.)

I reported my findings to the president of the company who said, "I pay them a day's pay for a day's work and that's all the damn motivation they need, or are going to get from me." In other words, the president himself was responsible for the problem! It was recommended to him that he read *Theory Z*, *In Search of Excellence* and other related books on motivation after failing to convince him of the importance of motivational programs. He, of course, was irritated by this suggestion.

It is not surprising to observe that this company has gone out of business.

Later it was found that this attitude was shared by many top executives, which I'm sure is partly responsible for the traditional conflict between unions and management in the USA. This subject is discussed in greater detail in Chapter 21.

SCHEDULING

The kind of scheduling system varies with the type of maintenance work. This is discussed in Chapter 16. If the scheduling system isn't intelligently designed, or if there is no system, productivity will be adversely affected.

This is true because proper scheduling is necessary to insure that manpower, instructions, material, tools and outages be brought together in optimum time relationships.

EMPLOYEE SKILLS

Quite obviously the skill of an employee affects his productivity. An employee who is skilled at a particular job can do it faster than one who is not and do it so that it won't immediately break down again. If there is a productivity problem with an employee, it is important to determine if it is a morale or motivational problem as opposed to a skill problem.

There are three action items required of each maintenance supervisor regarding this productivity factor:

- o Be aware of the skill level of all employees working for him.
- o Communicate this measurement to the maintenance manager.
- o Apply training effort commensurate with the individual's weakness.

A great part of skill weaknesses will be automatically taken care of by an apprentice program. However, specific training for particular individuals may at times be appropriate. Training is discussed in more detail in Chapter 22.

15

Planning

GENERAL

This section will discuss planning. Definitions of related functions follow.

The basic purposes of planning and scheduling maintenance work are:

- o To improve the productivity of maintenance labor
- o To perform work on schedule
- o To plan for future requirements before they are needed on a crisis basis

Planning is the determination of all elements required to perform a task in advance of the job start time.

Scheduling is the specific time phasing of the planned elements together with orders to proceed with the work, and follow up and status reporting on job progress.

Work accomplishment includes briefing the crew on the job elements, giving orders to proceed with the work, performing the work, job coordination and checking the quality of work in progress, and at completion.

Planning can be divided into three basic levels:

- o Daily job planning
- o Short range planning
- o Long range planning

Daily job planning concerns itself with the determination of all elements required to perform an individual task in advance of the start date.

Short range planning concerns itself with an annual plan of operation. This plan will specify how the maintenance force will operate during the coming year and will provide detailed plans for major overhauls and construction jobs. An important part of the short range plan will be vacation planning and plant shutdowns, if any.

Long range planning covers a period of five years or more and provides plans for future activities and for long range improvements to the maintenance function.

LONG RANGE PLANNING

Long range planning of maintenance requirements is closely allied with, and dependent on, long-range sales and production forecasts. Such planning is accomplished by staff planners who develop a long range program for the entire company. The planners, working with line executives of the operating divisions, outline, in effect, what is needed in the way of decisions today in order to reach certain goals five years from now. The management level of the staff long range planners is usually very high, reporting directly to the president or to a vice president. The long range plans are projected throughout management, and those affecting production influence the planning of the maintenance manager and others throughout the organization.

The primary purpose of long range planning is to keep the maintenance objectives, policies, and procedures updated in order that they will be in tune with the objectives of the company. This requires knowledge of sales and production forecasts and consideration of all the factors involved in long range production planning. In addition, maintenance planning requires projection of two particular factors which are very important to the maintenance organization.

- o Changes in maintenance equipment and facilities requirements.
- o Changes in production equipment due to obsolescence, increased mechanization, automation, and other technological improvements.

Thus, the planning of maintenance requirements for the future involves changes within the maintenance department as well as planning for the maintenance work to be done in support of production in the future.

Finally, the long range planning includes general plans and objectives for improving the operation of the maintenance function. This includes items such as training programs for craft workers, management development programs for future maintenance managers and foremen, methods improvement objectives, new maintenance requirements arising from proposed capital equipment acquisitions, equipment replacement plans, and long cycle equipment overhauls.

In summary, long range planning takes a look 5 to 10 years into the future and makes plans for the maintenance department to cope with coming events. Predicting what will happen in the future is done in several ways.

- o Studying the long range plan of production to determine what impact it will have on maintenance.
- o Extrapolative planning, that is, the extrapolating of present trends on into the future.
- o Entrepreneurial planning, that is, the prediction of maintenance impacts caused by changes in the external environment. This covers anything and everything outside of the company any place in the world that may impact the company's operations. Admittedly there will not be too much of this type of planning taking

place in the maintenance department but is important to top management.

Long range plans should be made out once per year by the maintenance manager. Progress towards long range plans should be reviewed on a quarterly basis. The maintenance manager should call upon members of his staff to submit ideas for the long range plan. These will not only provide valuable ideas to the maintenance manager, but will also make the staff feel that their input is wanted and therefore have a motivating effect.

Examples of long range plans are shown in Figures 3-3 and 3-4. Foremen developing long range plans will find the format in the examples helpful.

Foremen should avoid making long range plans that are too generalized and lacking in specifics, for example:

- "Improve the efficiency of the department," or
- "Improve communications," or
- "Reduce travel time."

This fault can be avoided by developing a plan that is as specific as possible and by listing all of the steps required to implement the plan complete with estimated completion dates. The maintenance manager should follow up on the due dates for milestones in the plans.

SHORT RANGE PLANNING

The short range plan is the annual plan of operation. It consists of the following:

- A general plan of how all labor should be distributed throughout the various departments under "normal" circumstances.
- The detailed planning and scheduling of major events taking place during the

Long Range Plan No. *1*

For

General Maintenance Foreman – *Smelter*

Date Initiated: *1-1-72*

Description: *Select, train and develop a replacement for position of General Maintenance Foreman – Smelter (in case incumbent leaves, dies or retires).*

Implementation Steps

No.	Step	Status/Remarks	Estimated Comp. Date	Actual Comp. Date
1	*With the help of the Personnel dept., select 3 candidates.*		*1-15-72*	
2	*Screen, interview and select one of 3 candidates.*		*2-15-72*	
3	*Study strengths and weaknesses of successful candidate. Develop plan for overcoming weaknesses.*		*3-15-72*	
4	*Rotate candidate through all smelter maintenance supervisory jobs including Planner/Scheduler's position.*		*1-1-74*	
5	*When sufficiently trained, use as acting General Foreman during vacations, sick leave, etc.*		—	

Figure 3-3: Long range plan–example No. 1.

Long Range Plan No. *2*

For

General Maintenance Foreman – *Mill*

Date Initiated: *1-1-72*

Description: *Reduce travel time for all maintenance units*

Implementation Steps

No.	Step	Status/Remarks	Estimated Comp. Date	Actual Comp. Date
1	*Develop procedure for measuring travel time*		*1-15-72*	*1-17-72*
2	*Brief foremen on procedure*		*2-15-72*	*2-1-72*
3	*Measure travel time (one week long study)*		*3-15-72*	*3-15-72*
4	*Meet with foremen and discuss study results*		*4-1-72*	*4-1-72*
5	*Write analysis and recommendations. Add recommendations to this plan.*	*Study concludes that problem will be solved by purchase of 4 golf carts*	*5-1-72*	*5-1-72*
6	*Purchase 4 Cushman gasoline engine driven golf carts*		*9-1-72*	
7	*Measure travel time (one week long study)*		*10-1-72*	

Figure 3-4: Long range plan–example No. 2.

year such as major overhauls, large construction jobs, major preventive maintenance tasks and vacations.

The first item shows the steady state situation. The second item shows the major perturbations to the normal schedule. These should be plotted on an annual Gantt chart of activities to give visibility.

The basic purpose of the short range plan is to provide adequate lead time to plan all elements of very large jobs and to describe the normal distribution of maintenance labor by work type. Lack of such a plan will inevitably lead to excessive downtime and costs due to unavailability of labor, materials, tools, equipment, access and instructions.

DAILY JOB PLANNING--GENERAL

Studies of typical maintenance departments often show that less than 25-30 percent of the day is spent on productive work, with more time spent on non-productive activities. Maintenance management has tried to find the answer to this basic problem of manpower utilization. The answer involves many factors and management tools.

One of the most effective tools used to prevent such waste of time and dollars is *daily job planning*.

The production of even the simplest of manufacturing parts requires a series of steps involving many people to carry the product from the original idea to design, selection of methods, tooling and materials, and coordination of purchasing, processing, finishing, and distribution. But the entire procedure for making a maintenance "product" is often carried out by one person, the maintenance craftsman, when, for example, he is sent out on a job to "see what the foreman wants in department 39 and take care of it." He designs the product, decides on methods and materials, secures tools and supplies, delivers and installs, and even decides what account to charge the job to! If we really want him to spend more time working at his trade, it seems only logical to provide a staff service to carry out some of these indirect operations. Stated simply, the way to avoid excess trips back and forth to the shop for manpower, tools, materials and equipment is to provide an accurate description of what is needed before the job is started. The way to decide how to do the work is to analyze the job *before* it starts. This is the primary function of maintenance job planning.

Daily job planning provides the following:

Job Instructions

How to do the job including drawings and specific-ations if required. It is important to understand that any design, plan or instruction omitted may result in lost time for maintenance labor several times greater than it would take to do complete planning. This is particulary true if the job requires a multi-man crew.

Material Required

The planner should provide a list of all materials required and write purchase requisitions for non-stock items.

Labor Required

An estimate of the labor required, broken down by trades and crew size.

Special Tools and Equipment

A list of special tools and equipment needed for the job, other than ordinary tools and equipment

Safety Instructions

A listing of unusual safety hazards involved in accomplishing the work and what to do about them.

Required Completion Date

The planner should arrive at the required completion date by estimating the following time requirements:

- o Design and/or planning time
- o Work order approval time
- o Material delivery time

- o Shop backlog time
- o Work accomplishment time

These should be added to the time the project was received to arrive at the required completion date.

This should then be compared with the requested completion date. If the calculated date is later than the requested date, the planner should negotiate with the requestor for a revision to the due date.

Work Orders

The planner should prepare necessary work orders.

DAILY JOB PLANNING--WHAT IS PLANNED?

All work should be planned in advance; even emergency breakdowns and trouble call work can be planned in advance as described later in this chapter.

DAILY JOB PLANNING--WHAT IS ITS PURPOSE?

The basic purpose of planning is to reduce costs. An unplanned job will almost always cost more due to inherent inefficiencies.

DAILY JOB PLANNING--WHO IS RESPONSIBLE?

First of all there should be only *one person* held responsible for planning *one job* regardless of the number of departments, trades, skills, men or engineering disciplines required to plan the job. Some *one* person has to:

- o Determine the components of the job. (In designing a battery charging room, he must know that an explosion proof exhaust fan is required for exhausting hydrogen gas. If he doesn't know this, the fan will be omitted.)
- o Coordinate between all people involved in planning the job.

- o Assume *total* responsibility for the design and/or planning of the job.

Without the assignment of one person as "project engineer" on a job, all of the above three points will become problems, with the result that portions of the job may be overlooked, there will be problems in coordination and if several people are involved without one person in charge, it will be impossible to hold one person accountable. Each person will say "Oh I thought John was responsible for that." Of course, these problems will not exist for a large number of jobs that would ordinarily involve only one planner.

The concept of having one person in charge of one project is simple and logical. However, it has been my experience that it is difficult to achieve unless the manager believes in the concept and perseveres, since it seems to be human nature to want to share the responsibility.

Job planning may be performed by any of the following people depending on the particular organization:

- o Maintenance foreman
- o Maintenance staff
- o Engineering staff
- o Planner scheduler

Furthermore, the maintenance manager may, at his discretion, decide to plan certain jobs himself.

Further guidance in who should perform planning and scheduling is given in the chart below:

LARGE PLANT	*MEDIUM SIZE PLANT*
Engineering-Engineering Dept. Planning-Engineers and/or Planners Scheduling-Scheduler	Engineering-Engineering Dept. Planning Scheduling (both by the Planner Scheduler)

SMALL PLANT	*VERY SMALL PLANT*
Engineering Planning Scheduling (all by the Planner/Scheduler)	Engineering Planning Scheduling (all by the Foreman)

DAILY JOB PLANNING--SCREENING

Screening is the process of questioning each maintenance job request in order to establish a firm economic basis as well as validity of need. The objective here is to put a dollar sign on the routine decisions regarding "Make or buy?," "Is it necessary?," and "How do we do the job?" Screening should include consideration of prefabrication instead of on-the-job assembly, purchased parts instead of parts made on the job or in the shop, replacement instead of repair, and contract service. In other words, this procedure helps to avoid overlooking a better or cheaper way to get the job done and in deciding if the work should be done at all.

Examples of excessive cost resulting from lack of screening can be found in almost any plant maintenance operation and include:

- Repair of fractional horsepower motors at more than the cost of new motors.
- Fabrication of workbenches and cabinets in a maintenance shop at several times the cost of commercial products.
- One-at-a-time machining of new standard gears that are available off the shelf.

DAILY JOB PLANNING--PREPARATION OF WORK ORDER PACKAGES

The planner, when making out a work order, should determine whether or not there is enough space on the work order to fully describe the work. If there is not, then the description should be short and refer to a supplementary document. For example:

CASE 1

"Furnish labor and material to fabricate and install a new control system for Mill No. 1 in accordance with drawing No. 12345, sheets 1 through 4 of 4;" or

CASE 2

"Furnish labor and material to rebuild Mill No. 1 in accordance with attached specification sheets 1 through 9 of 9." The specification sheet should have the same title as the work order and be identified by the work order number.

In other words the description for work should be COMPLETE, either by specifying everything on the work order or by referring to a supplementary document.

In Case 1, if there was a separate sheet for a bill of materials, then the description shown would be INCOMPLETE. It could be made complete by referencing a bill of materials on the work order or on the drawing (since the drawing is already referenced on the work order). The same applies to any other auxiliary document. The problem to be avoided with this procedure is the loss of any auxiliary document necessary to the work order without any indication on the work order that it exists.

DAILY JOB PLANNING--PROJECT PLANNER CONCEPT

If a job is poorly planned, it is usually for one of the following reasons:

- One person did not assume responsibility for the entire planning job, with the result that portions of planning were overlooked.
- Project planner was not qualified to plan the job.
- Project planner was careless in planning the job.
- Project planner didn't have the time to properly plan the job.

The project planner concept is simple. It merely states that for one job there shall be only *one* planner responsible. He must assume responsibility for coordinating the activities of any other people helping him in doing the planning.

The planner *must* determine if he is qualified to do the planning on a specific job. If he is not, he should disqualify himself and refer the job to someone who is qualified. Normally the maintenance manager should make this determination. Part of being qualified is having the knowledge of all the components that are necessary for a complete job. For example, in designing a boiler installation, he must know that a high pressure cutout is required. If he doesn't know this, it will be omitted.

Finally, if the planner doesn't have the time to plan a job, he must be aware of the consequences:

o If he holds the job in abeyance, what are the consequences of the delay?

o If he plans to put the job into work without planning, what is the extent of lost productivity?

If either one of these consequences is unsatisfactory, he should:

o Hold up another job and plan the job under scrutiny, or

o Arrange for someone else to plan the job.

The extent that a job is planned will depend on the type of work involved. The following paragraphs comment on the extent of the planning function by work type.

DAILY JOB PLANNING--EMERGENCIES

Previously, emergency work was defined as unplanned work and this is basically true. Nevertheless, there is a certain amount of pre-planning that can and *should be* performed. Examples of this are:

(a) Listing, by each foreman, of typical

emergencies that may take place in his area, giving emphasis to emergencies on critical equipment and emergencies involving a safety hazard.

(b) Writing of emergency procedures covering list (a) that cannot be handled by normal knowledge and skills of incumbent personnel within reasonable time frames.

(c) Training of personnel in procedures developed in (b); including, in especially critical situations, repetitive "Fire Drill" type of training.

(d) Preplanning of availability of materials either in stock in the warehouse or at the equipment for the purpose of reducing response time.

(e) Preplanning of availability of special tools, fixtures, and equipment for reducing response time. This should include consideration of design of special repair aids, especially on critical equipment. Consideration should be given to locating the tools, fixtures and other aids at the equipment.

DAILY JOB PLANNING--PREVENTIVE MAINTENANCE

Preventive maintenance work is by definition all preplanned.

DAILY JOB PLANNING--CONSTRUCTION, MODIFICATION AND REARRANGEMENT

All of this type of work should be completely preplanned.

DAILY JOB PLANNING--TROUBLE CALLS

Trouble call work or minor breakdowns is a type of work that is difficult to preplan. The obvious reason for this is that you don't know what to do until you arrive at the machine and diagnose the trouble. Further, it is frequently necessary to dismantle the machine and look inside it before you can diagnose the trouble.

Nevertheless, a certain amount of preplanning is possible. The following suggestions are made:

Troubleshooting Skills

Troubleshooting requires the highest order of maintenance know-how. Each maintenance unit should train and develop a crew of experts in this field.

Troubleshooting Aids

Certain items of equipment are easy to troubleshoot; others are complex and difficult. Trouble analysis charts should be made up and hung in plastic envelopes on the complex machines. These charts should follow the format

- o Trouble
- o Probable cause
- o Remedy

With these charts in his possession, the troubleshooting mechanic can quickly identify the trouble, its probable cause, and how to fix it.

Trouble analysis charts of this type are usually provided by the equipment manufacturer, but are of little value if hidden in some unknown location in a foreman's desk.

Troubleshooting Materials

Some preplanning of materials can be done. For example, it sometime pays to have a troubleshooting cart, or hand tote box, supplied with items typically used on trouble calls. Another suggestion is to store minor materials at the machine. For example: fuses, special nuts and bolts, packing, insurance spares or electrical brushes.

Troubleshooting Tools and Equipment

It goes without saying that mechanics should have

in their possession hand tools and instruments required for the job. However, thought should be given to other aids such as conveniently located ladders, hoists and cranes for lifting heavy parts and special tools, etc. In maintenance centers located remotely from the central shop, considera-tion should be given to creating subcenters for the purpose of storing needed tools and equipment that will reduce total travel time.

Training in Failure Modes

Operators and trouble shooting mechanics should be trained in all possible failure modes of critical equipment so that they will know how to respond to the failure. Lack of knowledge in these cases can be fatal (Chernoble disaster in Russia and the Three Mile Island incident).

Emergency Failures

Emergency failures on certain critical equipment such as boilers, generators, etc., must be antici-pated in advance by planning for such failures. For example, what procedure do you follow if the boiler feed pump stops? This is an emergency.

A list of all such possible emergency failures should be made.

Procedures should be written on what to do for each emergency failure that cannot be handled by "normal" skills.

Repetitive "fire drill" type training should be given to "appropriate" personnel in carrying out the procedure. The word appropriate is significant and must be carefully considered. For example, the immediate response for the boiler feed pump failure should be by the boiler plant operator. However, the follow on response should be by the trouble shooting mechanic due to the nature of their respective job descriptions and talents.

CASE STUDY 3D: The Smelter

While examining the records in the office directly alongside one of four huge copper smelters, I heard a loud explosion. When I came out of the office, everyone was running around like a chicken with his head chopped off. There was smoke all over the place and everyone looked scared. (I was also scared because no one seemed to know what they were doing leaving one without any confidence.) The explosion apparently came from the burner front, and the operators looked totally confused and were shouting for the office personnel to call the supervisor of instrumentation since he was the *only* one who really understood the smelter controls and how to operate the smelter when part of it failed as was the case during this incident. I considered the fact that the operators didn't know what to do was inexcusable since this was their job and that there really wasn't time to wait for the supervisor to arrive on the scene.

I reflected on the test I took for a chief marine engineer's license which lasted 5 days, 8 hours per day. Many of the questions asked related to how to continue operation in partial failure modes. For example, one of the questions was "what would you do if you heard loud muffled explosions coming from the boiler and at the same time saw the casing buckling in and out and what was the possible cause?" The answer was that this meant that the boiler was suffering from insufficient combustion air and that the boiler fan should be speeded up by opening up the steam throttle valve wider. In other words they wanted an exact, specific answer that indicated you knew what to do. The fact that a license was required certified that you possessed certain minimum qualifications. Unfortunately, such licensing and training practices are frequently not done in industry.

Managers must make the decision of which items of critical equipment by virtue of cost or safety should come under this type of thorough approach.

DAILY JOB PLANNING--MAJOR MAINTENANCE

Major maintenance jobs or major overhauls should all be preplanned. As a matter of fact, most major overhauls should be part of the short range plan outlined earlier in this chapter.

Major overhauls can be divided into two categories:

- o Known Work
- o Unknown Work

This can best be illustrated by examining the overhaul plan for a reciprocating pump:

Known Work

(a) Completely disassemble pump.

(b) Renew all broken nuts and bolts.

(c) Renew all gaskets, packing and piston rings.

(d) Clean all surfaces; remove rust and scale.

(e) Check all valve springs and renew broken or defective springs.

(f) Check all valves for proper seating.

(g) Paint all surfaces as required with paint spec XYZ.

(h) "Mike" cylinder for barrelshapeness and out of round.

(i) Check shaft, sleeves, and bushings for excessive wear or scoring.

Unknown Work

- o As a result of item (f) above, valves may have to be renewed and/or ground in.
- o As a result of item (h) above, cylinder may require reboring or renewal of liner.
- o As a result of item (i) above, shaft and sleeves may require renewal, building up and/or honing.

The point to be observed in this example is that a major portion of the overhaul work can and *should* be preplanned and can be estimated.

A critical item to be preplanned in major overhauls is

the coordination required with production to obtain a mutually agreed upon outage time which will allow all of the elements of planning to be accomplished and a reasonable amount of time to do the work. To do less is to invite increased costs and equipment downtime.

On very complex items of equipment, particularly where downtime for overhaul causes a production loss, it may pay to draw a "PERT" or "CPM" diagram for the overhaul to maximize coordination, and reduce confusion and lost time.

DAILY JOB PLANNING--SAFETY ITEMS

The safety department should conduct routine *scheduled* inspections of facilities and operations to insure compliance with company safety practices.

Violations to these practices shall be recorded by the safety engineer making the inspections.

The safety engineer should then prioritize each item in accordance with the priority system described in Chapter 9.

Finally, the safety engineer should make a list of prioritized safety violations. A separate list should be made for each foreman, showing only the items under his jurisdiction. The list should be sent to each foreman, and a copy to the maintenance manager with instructions to eliminate the safety hazards in priority order.

A priority system specially designed for safety items, as an alternate to the system described in Chapter 9, is shown below.

EMERGENCY--Priority 1

Safety items in this category constitute an existing and immediate threat to life and should be acted upon immediately. The safety engineer should have the *authority* and *responsibility* to shut down the operation if in his opinion it is necessary to protect life. Emergency items should be reported immediately to the maintenance manager by phone. Modification or new construction work required to eliminate the hazard should be covered by "after the fact" work orders.

URGENT--Priority 2

Safety items in this category constitute an imminent threat to life or limb and should be acted upon within a week. Items in this category should be approved by the maintenance manager. If items require maintenance support, the planner/-scheduler should make out and log the necessary work orders. The work orders should be signed by the maintenance manager.

NON-CRITICAL SAFETY--Priority 3

Safety items in this category should be accomplished within 3 months. Except for this, the items should be processed as described under Urgent items.

In the event the maintenance manager disagrees with a safety discrepancy, a meeting should be arranged by the manager with the safety engineer for the purpose of resolving differences in opinion. In the event the conflict cannot be resolved at this meeting, it should be resolved at the general manager's or president's level. The safety engineer should prepare a monthly list of delinquent safety items for submittal to the general manager with a copy sent to the maintenance manager.

The safety engineer should be responsible for following-up on all safety items until they are accomplished, or, in the case of contested items, until the issue is resolved by the general manager.

The planner/scheduler should be responsible for reporting the status of safety items to the safety engineer on a weekly basis. He should also be responsible for following-up on safety items requiring maintenance support to insure compliance with schedule requirements of the priority assigned.

The following case studies describe examples of various types of good and bad planning efforts:

CASE STUDY 3E: The Hose Rack

One day I got a call from the fire chief. He wanted to know where his hose rack was for drying hoses. He explained that all fire hoses that were used had to be dried, otherwise the water would rot them and they would burst the next time they were used. He said that he had ordered the racks one month ago.

I called the maintenance supervisor who said the racks were finished and delivered yesterday. I relayed this information to the fire chief.

He called me back in an hour and said that he couldn't find them.

I called the maintenance supervisor again who said that the racks were lying on the ground in the *back* of the firehouse directly by the rear door I relayed this information to the fire chief.

He called me back in 1/2 hour to announce that the racks were definitely *not* there. I told him I would be over with the maintenance supervisor shortly.

When all three of us arrived at the back of the firehouse it soon became apparent what was wrong. Instead of four 60'0" long drying racks, there were four 60 *inch* long miniature racks which, of course, were useless looking very much like they were designed for Mickey Mouse's fire station. It was found that the fire department planner had made an error in the sketch he drew up for maintenance.

Later on, I had difficulty convincing the maintenance planner that the fault was entirely *his*! I explained that we received requests from many people in the plant, most of whom were not engineers and that we constituted the engineering arm of the company. Once he (the planner) was assigned to the project, it was *his* responsibility to do the job right. He should have asked his client (the fire department) what the racks were for and he should have designed racks suitable for the job even without a sketch from the fire department.

CASE STUDY 3F: The Battery Charging Room

I might as well start off by saying that I refer to the people we serve (we, meaning the plant engineering and maintenance department) as clients. I picked this up from being in the consulting business for years. The reason I promote the use of this word to my subordinates is that it makes clear to them, by virtue of the consulting analogy, that we are a service organization and that we rise or fall in the company based upon our reputation with our *clients*! It certainly is a fact in the consulting business, but for some reason this fact of life escapes many in industry. Forgetting this fact usually leads to the destruction of the plant engineering and maintenance manager.

One of my clients called to complain about his new battery charging room not being ready and that every week we were putting the completion date back 2 weeks. He also said that he turned in the request 6 months ago. I was told by the project manager that the room was complete and ready for occupancy. I called the client and told him the good news.

One week later I got a call from the same client, who was so mad that he said he was thinking of dropping a bomb in my office. He said that the safety engineer visited the new facility and shut it down because it did not have an exhaust system for the hydrogen gas being emitted from the batteries. The safety engineer said the electrical system had to be explosion proof. This, of course, equated to long delivery times. Needless to say it took another two months to get the room operational.

I had a long talk with the project engineer who said that he had no idea that an explosion proof exhaust system wa required by the regulations. This case points out the importance of the supervisor selecting a project planner who is experienced in what he is about to design so that he at least knows all the elements of the total design. As an alternate to this the project planner should at least:

- o Discuss the design with his supervisor.
- o Consult with known authorities on the design inside or outside the company if he feels uncertain about his ability on the assigned project.

- o Do research in the library.
- o Call up vendors for advice (i.e., in this case, a battery manufacturer could have told him about the exhaust fan requirement).

Plant engineers are different from consulting engineers in that consultants are frequently expert in one narrow field and have many years' experience in that field, whereas plant engineers are frequently called upon to design things that they never have done before or have done only once. This can be dangerous. The assignments must be handled with common sense and a full understanding of the capabilities of personnel.

CASE STUDY 3G: Astronaut Rescue

At one point in the development of the Apollo Spacecraft we had a great tragedy. This was the death of three astronauts by fire. As you may recall they were on the ground and an ignition source started a fire in the command module where they were seated in a 100% oxygen environment. This caused them to be cremated in seconds.

Afterwards there was a tremendous redesign effort to eliminate combustibles in the command module. In addition, an astronaut rescue team was created for which I was given the responsibility. The team members were selected from the best men in the fire department who were willing to serve and who had rescue experience.

Concurrently with the command module redesign several different systems were designed to rescue the astronauts (in case of some malfunction) while they were strapped in their chairs inside the command module. The rescue operation consisted of the following steps:

- o Getting the rescue team to the module.
- o Extracting the astronauts by unbolting the cover and helping them out.
- o Escaping with them down one of the alternate escape mechanisms or systems.

Since there was so little time in an emergency situation, we conducted "fire drill" type exercises for each escape alternate to hone the rescue team into a smooth, efficient and fast operating team capable of performing all three steps within the very short allowable times.

The discipline described above has direct application in industrial plant maintenance. I used it for the first time in industry in overhauling a very large ball mill (there were 30 other identical machines). The mill was a critical machinery item in this plant and the general manager was complaining that the length of time spent on these overhauls was killing their productivity. My task was to reduce the overhaul time on the ball mills in the plant. I did this by doing the following things:

- Design a CPM (critical path method) network diagram with the help of the old timers who did the job many times before. This diagram not only showed what tasks were to be accomplished, but also the start and completion dates and the interrelationships between tasks. It also allowed you to "fool around" with the sequencing on paper which was much less expensive than doing it while the actual job was in progress. This method (CPM) is frequently used in construction but seldom in maintenance.
- Holding many meetings with the old timers to solicit ideas for cutting down overhaul time.
- Designed special jigs, tools, ladders, fixtures and material handling equipment to facilitate the removal and reinstallation of the very heavy parts. In some cases, these devices were permanently installed at the ball mills.
- Established a policy for making rebuilt accessories or parts available for immediate installation where appropriate. In some cases these parts were stored right at the mill.

- o Kitting materials prior to job start.
- o Obtained outage approvals prior to job start.
- o Arranged for the required skills and numbers of the various trades to be available, being careful to select people who were familiar with the work.
- o Arranged for a time study of the job in process to determine if the method, as depicted on the CPM, was most appropriate and to put down actual times. This group was charged with analyzing the job and making suggestions for improvement until such time as the method appeared to be perfect.
- o Arrange for the project manager of the task to perform a dry run as far as was possible, without the machine being opened up, so that the maintenance manager could judge his team's fitness and understanding of the project.
- o After the method was judged to be perfect, 2 other project managers were trained to be fully competent in the new method. This was done because there were 30 identical ball mills and to have more than one person available in case of multiple tasks in process at the same time or in case of absenteeism or vacations.

The moral to this case study is that we all should recognize that there are times when normal planning procedures should be superceded by the special or superhuman effort. The cases cited represent two cases where a superhuman effort was warranted:

- o When human life is at stake.
- o When large dollar losses might result from poor planning.

16

Scheduling

GENERAL

Scheduling is one of the most important tools used by management for insuring high labor productivity and the orderly accomplishment of tasks.

Scheduling arranges to have the planned elements available *prior* to the job start time.

Typical of the elements required to perform a job are:

- Labor
- Material
- Tools
- Equipment
- Equipment outage
- Safety lockouts
- Instructions

This section describes a recommended scheduling system. All of the required scheduling systems are described. Each individual organization may want to modify, add to or delete from these recommendations. Some person should be assigned to handle scheduling. This may be a foreman, a planner/-scheduler or a full-time scheduler depending on manpower availability and the size of the plant. It is quite common and it is recommended to have at least one person serving as a combination planner and scheduler.

SCHEDULING FUNCTIONS OF THE PLANNER/SCHEDULER (PS)

The Planner/Scheduler has the following scheduling responsibilities:

- o Allocating manpower
- o Managing capacity and workload fluctuations
- o Estimating
- o Scheduling of work
  - PM
  - Trouble calls
  - Major maintenance, construction, mods and rearrangement
  - Routine assignments (i.e., janitorial work)
  - Deferred maintenance
- o Managing the manpower pool
- o Control of backlog
- o Monitoring the flow of WO's

The planner/scheduler should be one of the most knowledgeable men in maintenance about operations in general, operating problems of the day, outages, emergencies, etc., so that he can effectively function as a coordinator between operations and maintenance.

Supervisors should be able to come to him and get answers to the following questions:

- o Where is labor assigned in general? This information comes from the monthly labor assignment sheet (see Figure 3-5).

	Hours per day
A. Unit Capacity	
10 men x 8 hours/day	80
Overtime	0
	80
B. Routine Assignments	
1. Fixed	
a. PM	8
b. Other	8
2. Variable-trouble calls	8
	24
C. Unavailable Time	
Absentees, vacations, sick leave, training, union business, etc.	8
D. Nonroutine Assignments	
Time available for special work orders A-B-C = D	48

• Each area foreman should prepare a separate labor assignment sheet for each trade.

Figure 3-5: Labor assignment sheet.

- o Total men missing because of absences, vacations, sick leave, transfers, etc.
- o Current emergencies
- o Location of all files and records
- o Location and availability of drawings and manufacturer's instructions
- o Impending and current outages
- o Work order status

ALLOCATION OF MANPOWER

The first step in the scheduling process is to prepare a labor assignment sheet for each of the area maintenance units. This is the first part of preparing the short range or annual plan. These sheets, which show how labor is distributed by job type, should be updated monthly, if required.

The assignment sheets are not meant to show the exact assignments on a daily basis, but rather are used to show the basic manpower allocation in general and to serve as a point of departure in making the inevitable daily shifts in manpower required by changes in manpower capacity and work load changes. The labor assignment sheet is shown in Figure 3-5. An explanation of this sheet follows:

A--Unit Capacity: This is the number of permanently assigned personnel times 8 hours per day. Pooled personnel should not be included in this figure. In addition, overtime capacity should be added only if it is planned for the entire month.

B--Routine Assignments: "Routine Assignments" should be divided into two categories:

Fixed Assignments of Known Magnitude

> *Preventive Maintenance Inspection Route Work*--Foremen should go on the route at least once to measure the amount of time it takes. Once this is done, this daily assignment is of known magnitude. Manpower should be allotted to it and should be left undisturbed by other work.
>
> *PMI's (Preventive Maintenance Instructions)*--Each week the computer prints the week's schedule of preventive maintenance tasks derived from preventive maintenance instructions. This manpower is known beforehand and should be set aside and left undisturbed.
>
> *Other Fixed Assignments*--There will be other fixed assignments of known magnitude such as oiling, lamp replacement, and janitorial work, etc. This manpower can be calculated and set aside.

Routine Assignments of Unknown Magnitude--Examples of this category are:

Minor maintenance

Operating foremen surfaced preventive maintenance

Emergencies

Trouble calls

The amount of manpower required for this type of work can be estimated by averaging past experience shown on the blanket work orders.

C--Unavailable Time: This it the average time spent by the crew on absenteeism, vacations, sick leave, coffee time and union business, etc. This figure can be obtained from computer records, past experience and knowledge of the current situation.

D--Nonroutine Assignments: This item is the amount of labor available for nonroutine assignments, and is equal to A minus B minus C. This category of work comprises practically all of the preplanned workload and is the major scheduling task of the planner/scheduler. One of the main purposes of the labor assignment sheet is to enable the computation of this quantity.

MANAGING THE MANPOWER POOL

There may be several separate area maintenance groups, each headed by a foreman. During any one day the workloads of each of thee groups will vary. Variations may be caused by breakdowns, changes in priorities, jobs, inability to obtain outages, and a wide variety of other reasons.

Furthermore, the crew capacity available to perform the work will also vary for the following reasons:

- Absences
- Vacations
- Sick leave
- Transfers to emergency jobs

- o Transfers to higher priority jobs
- o Inability to shut down equipment for servicing (prevents use of labor)
- o Terminations
- o Union business
- o Inability to obtain material or equipment (prevents use of labor)

It is the responsibility of the planner/scheduler to be aware of these variations and to make adjustments to balance load and capacity.

Adjustments can be made by one or more of the following methods:

Insufficient Capacity/Excess Load

- o Overtime
- o Outside contracting
- o Delay, reduce or cancel work
- o Inter-unit transfers
- o Allocation of manpower pool labor

Excess Capacity/Insufficient Load

- o Schedule more backlog work
- o Schedule deferred maintenance tasks.
- o Preschedule selected preventive maintenance tasks. (The planner/scheduler can add workload by scanning next week's preventive maintenance tasks, particularly quarterlys, semi-annuals, and annuals. It will not hurt the preventive maintenance program to schedule an annual overhaul, for example, one week sooner.)

IT MUST BE ASSUMED THAT THERE WILL BE TIMES WHEN ONE MAINTENANCE UNIT IS OVERLOADED WITH WORK AND ANOTHER IS UNDER-LOADED. IF NOTHING IS DONE ABOUT THE SITUATION, WE CAN EXPECT THAT THE UNDER-

LOADED UNIT WILL WORK AT LESS THAN PEAK EFFICIENCY. FURTHERMORE, THE OVERLOADED UNIT MAY HAVE TO DEFER WORK THAT IS CRITICAL TO PRODUCTION. CLEARLY EITHER SITUATION IS TO BE AVOIDED. THE SOLUTION TO THE PROBLEM IS TO PROVIDE A MANPOWER LABOR POOL.

The question is, how big should the pool be?

The answer to this question can be obtained by first determining the size of each area maintenance unit. Each area maintenance unit should have enough manpower to take care of the average labor required for routine assignments and unavailable time (items B & C of Figure 3-5).

In addition there should be labor set aside in an amount equal to the *average* of the variable nonroutine assignment workload (item D of Figure 3-5).

The next step is to divide item D (non-routine assignments) into two parts, i.e.:

- o Permanently assigned to the area maintenance unit.
- o Assigned to area maintenance unit but allocated to the manpower pool.

The method used to divide item D is illustrated in Figure 3-6 and explained below:

The difference between the average and the average minimum nonroutine labor quantity will be the particular area maintenance unit's contribution to the manpower pool. This amount of manpower will routinely remain in its home area maintenance unit and be assigned elsewhere only when the planner/scheduler decides there is a need. In other words, this isn't a pool of men in the traditional sense that normally gathers in one place awaiting assignment. Stated differently this arrangement constitutes a preplanned agreement between the maintenance manager, his foreman and the planner scheduler that certain men are permanently assigned to a unit and others are subject to redistribution to other units on a daily basis.

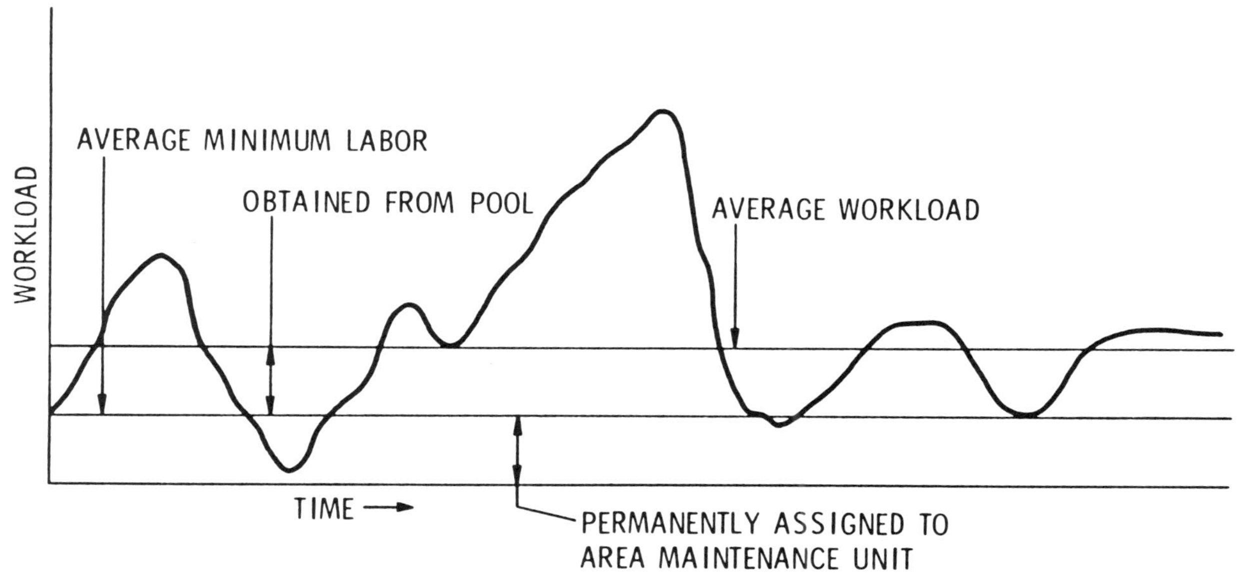

Since there will be times when one maintenance unit is overloaded and another underloaded, a manpower labor pool should be provided for non-routine labor assignments to absorb the brunt of workload fluctuations.

Figure 3-6: Manpower pool determination.

With this method of determining the area maintenance crew size, there should *always* be a sufficient workload for the fixed staff. In addition, there will be a pool large enough to handle the majority of labor fluctuations without borrowing continuously from the fixed area maintenance staff.

HOW PREVENTIVE MAINTENANCE INSTRUCTIONS ARE SCHEDULED

The scheduling function for preventive maintenance instructions is performed by the computer as described in the section on preventive maintenance. All preventive maintenance instruction cards should be distributed by the planner/scheduler each Friday afternoon to the appropriate area foremen. The area foremen should have all of the following week in which to accomplish the preventive maintenance instructions. There will be little involvement of the planner/scheduler other than to coordinate outages if required.

As a result of the load balancing effort of the system, the manpower required for preventive maintenance instructions should be relatively constant and even. Consequently, a fixed staff should be set aside for this function. In the case of insufficient preventive maintenance instruction workload, men can be assigned to backlog work. In the case of a momentary peak caused by perhaps a large annual overhaul, men can be borrowed from the manpower pool.

HOW TROUBLE CALLS ARE SCHEDULED

As previously stated, the average manpower expended on trouble calls should be determined and set aside in each area maintenance unit.

In areas where the workload is small, the fixed crew can work alternately on preventive maintenance instructions, trouble calls, fixed assignments and daily Inspection Route Sheet Work. The detailed every day scheduling of men to this type work should be handled by the area foremen rather than the planner/scheduler. In this regard the planner/scheduler will be involved only in major workload/capacity fluctuations.

Trouble calls should be received at a "trouble call desk," which should have a separate telephone line set aside just for this purpose. The trouble call desk may be manned by a clerk, who will receive and dispatch trouble calls to the area foreman.

Trouble calls which appear to be over the dollar or hour limit of the appropriate trouble call blanket work order, should be referred by the clerk to the planner/scheduler for input into the work order system.

It should be the responsibility of the foremen to periodically audit the performance and work assignments of the trouble call mechanics to assure that they are:

- Working with reasonable productivity
- Working on approved types of work
- Handling an appropriate share of the trouble call workload

The planner/scheduler should analyze the job type and quantity of trouble calls. The following information should be obtained:

- Highly repetitive trouble call areas or types
- Average time per trouble call per type
- Numbers of calls per month
- Average hours per job type by employee

All of these statistics can be useful in surfacing trouble spots, poor performance and in scheduling of work.

HOW FIXED ASSIGNMENTS ARE SCHEDULED

The scheduling of fixed assignments is a simple task accomplished as follows:

1. Define task
2. Estimate manpower required
3. Assign estimated manpower to task

4. Periodically audit performance to task including:

    - o Task completion
    - o Work quality
    - o Conformance to estimate

Examples of fixed assignments are:

- o Janitor's work
- o Daily inspection route sheet preventive maintenance work

Using the daily inspection route sheet preventive maintenance work as an example, the planning and scheduling task would proceed as follows:

1. Define task--This is done by filling out the daily inspection route sheet form.
2. Estimate manpower required--This can be done simply by walking through the task with a competent mechanic working at a reasonable rate to see how long it takes him. In the case of a janitor, the task can be estimated from detailed work standards.
3. Assign estimated manpower to task.
4. Periodically audit performance--The form filled out by the mechanic tells what he was supposed to have performed; the foreman should periodically audit on an after-the-fact basis if he did what he signed off as having completed. At the same time, the foreman should check the actual vs. estimated time and the quality of work. In the case of the janitor, the foreman can periodically check the janitor's performance to task, quality and estimate. He should also take note of customer complaints. Lack of these auditing steps invites poor performance when the maintenance worker discovers

that no one ever checks to see whether or not he is doing his job.

The point to be remembered on fixed assignments is that if the 4 steps are not followed you can't tell:

- o If the task is being accomplished
- o If the quality of work is satisfactory
- o If the productivity is adequate
- o If the correct amount of labor is assigned

CASE STUDY 3H: The Janitors

The president of a very large plant asked me to study the janitorial operation which consisted of 450 janitors. He was dissatisfied with the average cost of janitorial services measured in cents per square foot which was much higher than it would cost him to contract out. Nevertheless he was constrained by the union agreement not to contract out the services.

I toured the facilities and could see that the plant and rest rooms were a mess. Investigation revealed that there were no janitorial standards and that the job methods were determined by each individual janitor according to his experience and motivation. I could see immediately that this was *the* problem.

I then held a meeting with the president and the maintenance manager and told them of my findings. I said that this type of error was quite common. People looked upon the janitorial task as being a very simple, mundane one, requiring no great talent and therefore permitted the planning of the tasks to be done at the lowest and frequently least qualified level.

There might be an excuse for this in a plant that has one janitor. In this case you would just look for a very qualified janitor and let him do his thing! However, in the case cited, we have a janitorial operation costing over $10,000,000. *THIS IS BIG BUSINESS AND WARRANTS PLANNING DONE BY A VERY, VERY ASTUTE PERSON!*

The moral to this case study is transparent: large dollar expenditures demand the best brains regardless of the technical complexities involved.

I was happy to get the sizable contract to develop a more efficient janitorial system for this plant.

HOW PLANNED JOBS ARE SCHEDULED

The scheduling of planned work orders will be the major scheduling responsibility of the planner/scheduler. It consists principally of major repairs, construction, modification and rearrangement work.

In order to systematize and clarify the scheduling process a planner/scheduler scheduling system should be installed near the planner/scheduler's desk. There are many types of systems that may be used. A very simple system which can be used is shown in Figure 3-7. This system incorporates all of the elements required in a scheduling system. The device is simply a place to visibly file work orders as they proceed through the various scheduling steps. Anything that serves the same purpose will do.

Referring to Figure 3-7, the scheduling process can be shown by describing the operations taking place as the work orders are manually moved through each of the 7 slots in the device.

Step 1--Incoming Work Orders

Upon initiation of the work order, all copies should be placed in this box. The work order at this stage should have all information called for except the planning information.

The planner/scheduler then decides whether to defer the work, in which case it should be filed in a deferred work file, or to process it for accomplishment.

If the planner/scheduler decides to process the work order, he then transfers all copies to the appropriate slot in box 2 depending upon who he decides should do the planning. The choices he has are:

- o Maintenance staff (this includes the planner/scheduler, the maintenance manager or other staff personnel)
- o Maintenance foreman
- o Engineering personnel

After the decision is made as to who should do the planning, the name of the selected planner should be written on the work order. At this point, the work order should be opened in the accounting and/or computer system.

All copies of the work order should remain in box 2 during the planning stage, except the copy given to the planner.

Step 2--Awaiting Planning

After being assigned to the planning task, the planner should accomplish the planning.

Upon completion of the planning task, the planner should enter onto the work order any missing information including the estimate. If necessary, he should rewrite the work requested and job title boxes if the existing descriptions aren't meaningful. The planner may also change the priority upon approval of the planner/scheduler, if warranted.

The planner should attach any auxiliary documents such as the bill of materials, specifications, or drawings to the work order.

Step 3--Awaiting Material and/or Outage

Upon completion of the planning process the planner/-scheduler shall:

- o Transfer copies of the work order to box 3.
- o Audit the completeness and accuracy of the work order information including coding.
- o Send a copy of the bill of materials and outside purchase requisitions to the storeroom for performance of the kitting function.

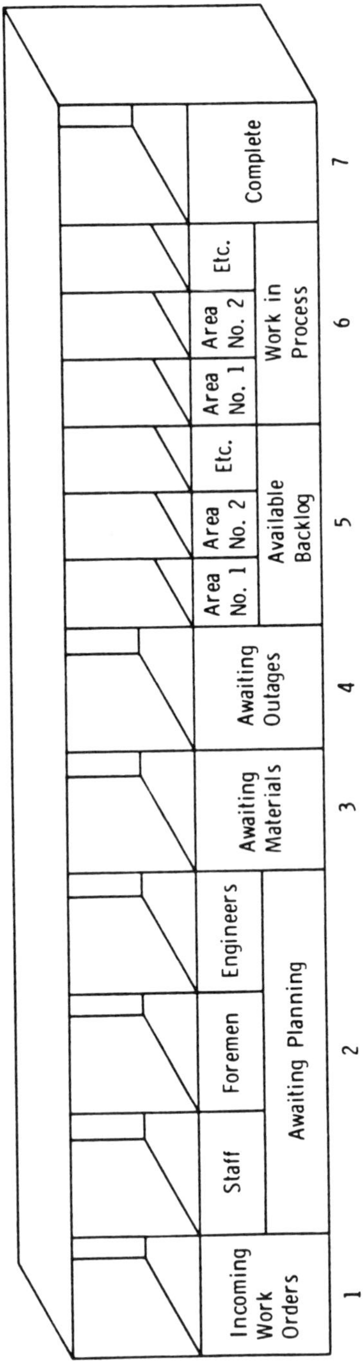

Figure 3-7: Work order scheduling aid.

Upon receipt of the bill of materials and the outside purchase requisitions the storeroom should proceed to kit the materials.

The storeroom has the responsibility of completing the kit by the due date established on the bill of materials.

If the storeroom discovers that the due date is going to be missed, they should immediately call the planner/scheduler, so that he can take corrective or alternate action with the help of the job planner.

Step 4--Awaiting Outage

If an equipment outage is required, the planner/scheduler should, upon completion of the material kit, transfer copies of the work order to box 4 and firm up arrangements for the outage called for in the job plan. Ordinarily this will already have been arranged and the work order will be put in slot 5.

Step 5--Available Backlog

Upon completion of the material kit, and obtaining of an equipment outage, if required, copies of the work order should be placed in the appropriate slot of box 5, depending on the area maintenance unit.

These work orders represent available backlog of work.

One of the prime purposes of the planning and scheduling function is to have an adequate "Available Backlog." This is one of the biggest steps forward in achieving high productivity. An empty box 5 means one or both of two things:

- There are too many people on the staff (not enough work).
- Some part of the planning and scheduling function is malfunctioning.

Step 6--Work In Process

Once per week, on Friday afternoon, the planner/scheduler should select, for each area maintenance unit, a week's worth of work from the available backlog with careful consideration to the job priority rating.

Upon selection of the workload, the planner/scheduler should send the work order and auxiliary documents to the appropriate foreman. One copy of the work order should be filed by the planner/scheduler in the appropriate "Work in Process" slot for follow up purposes.

The foreman then has the entire next week in which to complete the week's planned workload subject to:

- o Priorities
- o Due dates
- o Prearranged outages

Step 7 - Complete

When the work is complete, the planner-scheduler should process the work order through accounting for completion.

BACKLOG CONTROL

The planner/scheduler should maintain a record of backlog similar to that shown in Figure 3-8.

These graphs will be useful to plan vacations, to facilitate the work planning effort and to make decisions on manpower including the need to hire, fire or alter the numbers in each trade.

In most plants which have their own computer, this information can easily be made and updated as frequently as desired by the computer in tabular or graphic form.

Figure 3-9, Managing Backlog, shows my recommended policy for what to do about too large or small a backlog as far as hiring or firing is concerned. As you can see from the chart, a policy must be determined on what level of backlog you must reach to kick off a hiring or firing decision. A separate chart should be kept for each trade.

MANAGING CAPACITY AND WORKLOAD FLUCTUATIONS

The workload in any maintenance unit can be subdivided into the following fixed and variable categories:

Workload Types

FIXED

- Preventive maintenance
- Other fixed assignments

VARIABLE

- Minor repairs (trouble calls)
- Major repairs
- Construction, modification, and rearrangement work

Workload Fluctuations

Workload can fluctuate for the following reasons:

- Changes in priority
- Variable nature of work, i.e.:
    - Breakdowns
    - Trouble calls
    - Major or minor breakdowns
    - Construction, modifications or rearrangements
    - New jobs
- Inability to obtain outages
- Other reasons

Labor Capacity Fluctuations

The labor capacity available to apply to any of these fixed and variable loads is variable. There are many reasons for this, including:

- Absenteeism, vacations, and sick leave
- Labor transferred to emergency jobs
- Labor transferred to higher priority jobs
- Inability to obtain material or equipment

There should be little, if any, backlog of PM or minor trouble call work. Backlog should be primarily major maintenance and construction work.

The planner/scheduler should maintain a graph of the backlog of work on a weekly basis for each craft. This information will be helpful in scheduling vacations, planning work, and determining allocation of manpower.

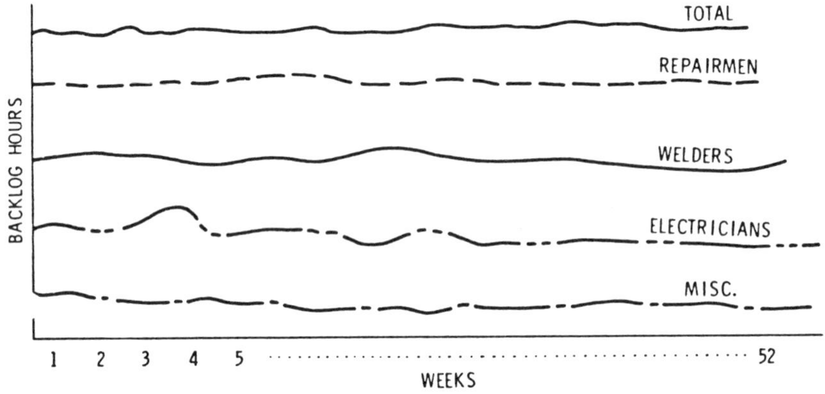

This chart is also used to:

- determine when to hire or layoff personnel
- interchange trades when allowed
- change labor mix
- determine allocation of deferred work or to defer work

Figure 3-8: Backlog of work.

- o Inability to obtain outage
- o Terminations
- o Union business
- o Variable productivity
- o Inability to obtain tools and equipment

In other words, the labor capacity to meet the needs of any particular load is dynamic and constantly changing. For this reason, it must be constantly monitored and adjusted. The following suggestions are offered to systematize and simplify one of the most difficult problems in maintenance management--the continuous balancing of load and capacity.

- o First of all, each foreman should decide to perform all PM work on schedule. Normally PM work is a fairly constant workload, so a fixed number of workers

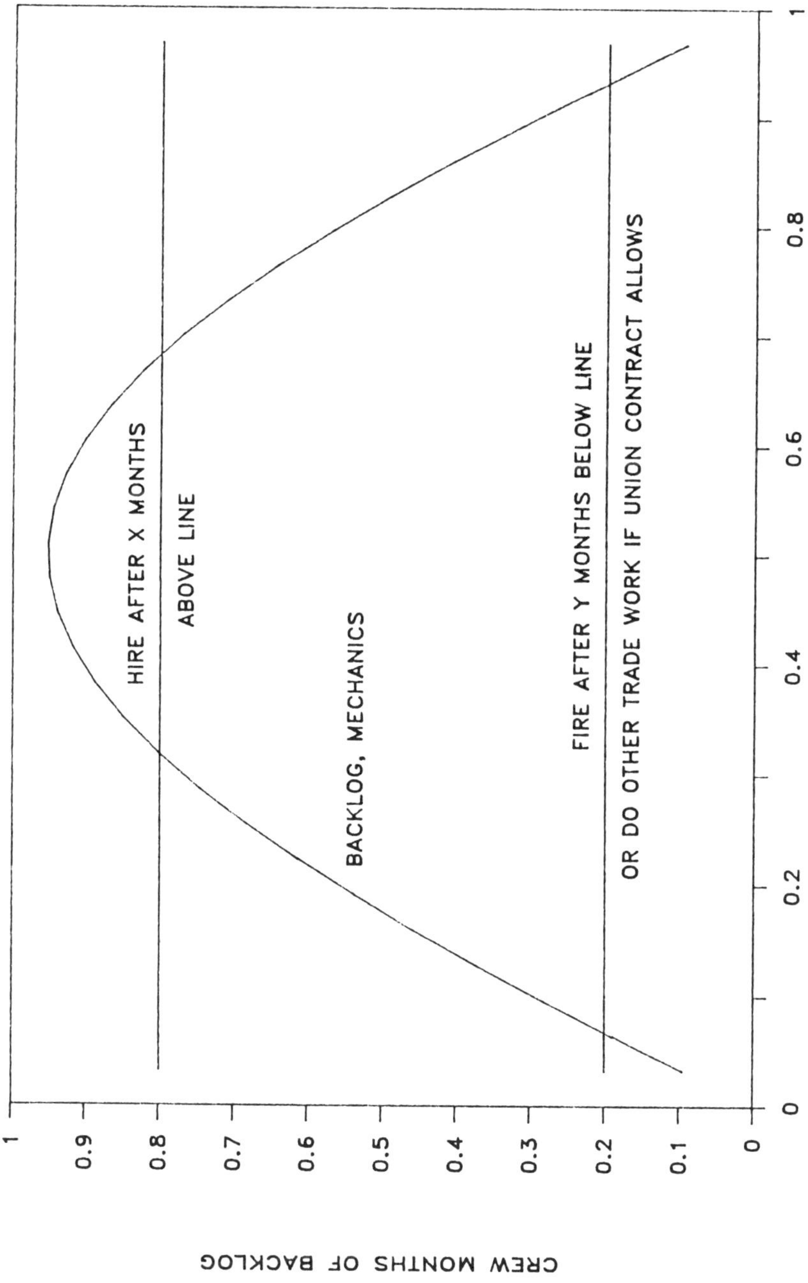

Figure 3-9: Managing backlog.

can be set aside to handle the load. The only variation then will be from vacations, sick leave, and absenteeism. This should be made up as it occurs. There will of course be a temptation to take people away from PM to handle emergencies, however, this should be avoided since it was probably lack of PM that caused the emergency in the first place.

- Concerning fixed assignments, decreased labor capacity caused by absences, etc., should be decided upon a case by case basis. Certain fixed assignments must be manned all the time at design strength. An example of this might be a particular oiling route. In the case of an electrician changing light bulbs, on the other hand, the assignment may be dropped for one or two days.

- Concerning construction, modification, rearrangement and major maintenance work, workload variations should be taken care of in two steps:

  - As previously stated the planner/scheduler should make weekly allotments of the manpower pool to handle workload variations by area and priority.

  - At the daily maintenance meeting the planner/scheduler should make adjustments to the manpower pool allocations depending upon the problems of the day.

- Finally, the other variable load in area maintenance will be minor repairs, or trouble calls. With this type of work, since the average job is of short duration and since production downtime is frequently involved, it is desirable to have no backlog. This work should be completed *within one day.* This, however, creates a problem. Without a backlog, input must equal output. Since the workload of minor repairs is variable, we would have to have a variable work force in order to achieve this. This end may be achieved in the following ways:

- Working overtime
- Borrowing men from the manpower pool

In summary, it is the intention to have the manpower pool absorb the major brunt of load fluctuations because overtime is more expensive and outside help is usually not available at short notice.

This section on managing capacity/workload fluctuations can be condensed into the following tables:

MANAGING CAPACITY/WORKLOAD FLUCTUATIONS

Insufficient Capacity/Excess Load

- Overtime
- Outside contracting
- Allocation of manpower from pool
- Delay, reduce or cancel work
- Inter unit transfers

Excess Capacity/Insufficient Load

- Schedule more backlog work.
- Schedule deferred maintenance tasks.
- Preschedule selected PM tasks.

Capacity Workload Fluctuation Policies

- Perform PM on schedule - set aside fixed crew.

- o Other fixed assignments - some critical, i.e., oiling; others non-critical, i.e., changing light bulbs. Sacrifice non-critical to do critical.
- o Minor trouble calls should have no backlog, finish all same day. The reason for this policy is to avoid being swamped with a backlog of thousands of small jobs that can be completed by a small allocation of your total manpower. (This is the magic tool at work, "Don't let yourself be swamped by trivia when it can be handled and taken off your mind by a small crew.") This is also a big source of criticism. "I've had a 5 minute job setting in their backlog for 3 months!," is what your clients might say about you.
- o Capacity/workload fluctuations in all other work types should be handled as previously described.

DEFERRED MAINTENANCE

From time to time certain tasks originated from any source may be deferred for future accomplishment. Examples of these tasks may be:

- o Repainting of offices
- o Cleaning of a cooling tower
- o Overhaul of a truck engine
- o Replacement of a corroded pipe line

Possible reasons for deferring this work may be:

- o Justification is questionable
- o Heavy construction workload
- o Large backlog of more critical tasks

Most preventive maintenance tasks, for example, may be deferred without immediate impact on operations. However, an indefinite deferment may eventually cause damage to facilities and may result in lost production time. Items of this nature, which are judged to be worthy of future accomplishment, should be filed in a "Deferred Maintenance File" together with a rough estimate on each deferred item.

This file should be used for "fill-in" work when there is excess capacity. However, deferred work should be approved by the maintenance manager prior to accomplishment.

"Deferred" work should *not* be added to the normal maintenance backlog report but should be reported separately.

PROJECT SUPERVISOR CONCEPT

The project supervisor concept is the follow-on to the project planner concept. The project planner is the *one* person responsible for planning a job. The project supervisor is the *one* person responsible for accomplishing the work in accordance with the plan, regardless of the number of maintenance units involved in the job.

A project supervisor should be designated for each construction, modification, rearrangement, or major maintenance job. He should be selected by the maintenance manager. His name should be put on the work order. He should be held responsible for the successful completion of the total job regardless of the number of other maintenance units involved.

The project supervisor should be responsible for coordinating the work of other units and outside contractors with the job schedule; and for delegating responsibility to an acting project supervisor when he is not on the job by virtue of being absent, working at night or weekends or other reasons. This responsibility includes the absolute necessity of adequately briefing the acting supervisor on all elements of the job including unusual and safety problems.

PURPOSES AND USE OF THE WORK ORDER ESTIMATE

Formal estimating should be performed on all non-routine work orders. The main purposes of estimating are:

- o To facilitate scheduling. The estimate makes it possible to load each area with a week's worth of work.
- o To provide a basis for checking job productivity. The estimated and actual job costs will show up on the completed work order print out. Jobs that were performed significantly above or below the estimate should be investigated for method problems, incorrect estimating or other reasons.
- o To provide a basis for work order approval. No one wants to approve a job without knowing what it will cost.
- o For backlog control. The summation of each work order estimate provides a measure of the size of the backlog. Backlog control would be impossible without a job estimate.
- o To help in controlling low productivity caused by incorrect crew sizing.

Consider how long a garage owner would remain in business, if when a customer asks how much a car repair will cost, he replied, "I don't know, whatever it costs when the job is finished, that's what it will cost you!" He must, of course, give the customer an estimate so that the customer can decide whether or not to proceed.

17

Work Accomplishment

JOB FAMILIARIZATION

Job familiarization refers to the foreman becoming thoroughly familiar with a job that he is going to assign to maintenance workers. This function can be divided into three categories:

- *Jobs that he has planned himself*--In this case, it is not necessary to become familiar with the job since he is the one that conceived it.
- *Trouble calls*--In most cases, it will be sufficient only for the foreman to be aware of the trouble call. The mechanic will handle these jobs by himself at the discretion of the foreman. In cases of critical breakdowns, the foreman will undoubtedly elect to investigate the problem personally.
- *Jobs planned by others*--These jobs, which will be mainly major repairs construction, modification, and rearrangement jobs, are the principal ones requiring familiarization on the part of the foreman. He must check the completeness of the planning:
  - Drawings

- Bill of materials
- Instructions
- Specifications
- Etc.

If the foreman doesn't thoroughly understand the package, he will not be able to properly and completely brief his men and they will have to obtain the missing information themselves at the expense of crew productivity. If the foreman doesn't understand the package, he should obtain clarification from the planner.

INSTRUCTIONS TO CRAFTSMEN

Craftsmen should receive instructions from the foreman prior to proceeding with maintenance work. The instructions will vary by work type i.e.:

- o Fixed assignments
- o Trouble calls
- o Major construction and maintenance jobs

Fixed Assignments

Examples of fixed assignments might be the performance of a preventive maintenance inspection route or a janitor cleaning offices. Since in the later case the men will be performing the SAME TASKS EVERY DAY, proper and thorough instruction by the foreman to the craftsman is MORE IMPORTANT THAN FOR ANY OTHER WORK TYPE. Frequently this fact is not recognized because of the low skill level of this work type. The foreman should also:

- o Audit performance to his instruction periodically.
- o Be on the constant lookout for improved methods.

The practice of allowing craftsmen alone to break in new workers on a fixed assignment is to be avoided since they are not responsible for worker productivity. Craftsmen may be used to assist in breaking in new employees after the foreman has given the first indoctrination.

Trouble Calls

Foremen should review trouble calls to determine whether or not they want to investigate the job and give instructions to the craftsman, or simply let the craftsman handle the trouble call completely by himself.

Foremen should recognize that trouble call maintenance requires the highest order of maintenance skill, consequently, foremen should set aside their most skilled journeymen for this type of work. When this is done, and the foreman is fully satisfied with the competence of his trouble call staff, he can then let them handle trouble calls completely on their own without instructions.

It is understood, however, that foremen will go to the scene of a serious production stoppage or safety incident and personally direct the repair effort.

Major Construction and Maintenance Jobs

The greatest effort in giving instructions to the crew takes place in the case of major construction and major maintenance jobs. The following statement bears repeating!

> IN ORDER FOR A MAINTENANCE TASK TO BE PERFORMED EFFICIENTLY AND ECONOMICALLY, MANPOWER, INSTRUCTIONS, MATERIALS, TOOLS AND EQUIPMENT MUST BE AVAILABLE PRIOR TO THE JOB START TIME.

The instructions to the crew should include:

- Technical and schedule aspects
- Discussion on material availability
- Discussion on special tools availability
- Discussion on special equipment availability
- Safety
- Outages and coordination with production
- Portions of work done by others and time phasing thereof

- o Schedule commitments
- o Lockouts

SAFETY

FOREMEN ARE RESPONSIBLE FOR THE SAFETY OF THEIR MEN!

The foreman's concern over the safety of his crew is necessary to protect them from injury or death and because a serious safety incident may cause:

- o Production loss
- o Low morale
- o Low labor productivity
- o Labor relation problems

Foremen should work with the safety department in eliminating safety hazards.

Secondly, foremen should discuss safety when briefing the men on jobs and alert them to unusual hazards of a particular job and what he wants them to do to conduct the work safely.

Safety engineers will, when investigating all accidents, determine if the men have been briefed on unusual safety hazards for the specific job they were working on when the incident occurred.

Foremen should see to it that certain equipment items necessary for the safe conduct of a job are available to the men. Examples include:

- o Protective clothing
- o Safety glasses
- o Hard hats
- o Safety shoes
- o Breathing apparatus

- Masks
- Safe ladders
- Spark proof equipment in explosive atmospheres
- Etc.

At Kennedy Space Center, which had many hazardous working environments, a green stamp, which was put on the work order folder, was designed. The stamp read, "I have briefed my crew on the safety aspects of this job" (followed by a space for the foreman's signature). The rule on non-compliance to this procedure was a week off the first time the foreman forgot to sign the stamp, and firing the second time. This also applied if the foreman signed the stamp and was later found that he didn't brief his men. This severe policy was made necessary as a result of several serious accidents.

PLANT CLEANLINESS

Plant cleanliness is an important responsibility for everyone. An unclean plant may cause one or more of the following problems:

Low Morale---Nobody likes to work in an unclean environment.

Production Loss--Ore spillage, for example, may result in damage to bearings, belts, skirts, etc., and eventually cause a shutdown.

Health Hazard--Uncollected fumes, dust from spillages, etc., may result in medical problems.

Safety Hazards--Fuel or oil leaks may result in slipping and falling.

Fire Hazards--Uncollected trash, papers, oily rags, or other combustibles may create a fire hazard.

Low Productivity--An unclean plant creates an atmosphere of disorderliness and confusion and may lead to low productivity.

For these reason it is important that each maintenance

foreman makes his contribution to keeping the plant neat, clean and orderly and that he invite his men to do the same.

He should do this by developing a plan. The following suggestions are offered:

- Shop orderliness--Foremen should outfit themselves with whatever bins, cabinets, tool racks, etc., necessary to insure an orderly shop arrangement.
- Workers should be instructed to keep shop neat and clean.
- Workers should be instructed to clean up after each job, i.e., return tools and equipment to shop and clean up dirt, trash and debris generated during job.
- Foremen should evaluate the problems or damage to equipment caused by uncleanliness, spillages, etc., and coordinate with production on eliminating the problem. If foremen have difficulty in solving this problem, they should alert the maintenance manager to the problem. The manager should then try to resolve the problem with the general production superintendent.

FEEDBACK INFORMATION

The foremen are directly responsible for the accuracy and completeness of feedback information concerning maintenance failures.

The subject is covered in detail in Part 2, however, the following points bear repeating:

- The primary *source* of feedback information is the maintenance craftsman who observes specifically what was wrong when repairing equipment. He should enter this information on the Work Order.

- The primary checking for accuracy and completeness of feedback entries is the responsibility of the maintenance foreman.

MAINTENANCE ANALYSIS

Maintenance foremen are responsible for analyzing and correcting repetitive failures on equipment that is under their authority.

Foremen should solicit the aid of maintenance craftsmen in helping him with this function by asking them to keep their eyes open in the field and to be sensitive to problems they personally witness.

LABOR AND MATERIAL REPORTING

Foremen should do whatever is necessary to support the accounting system designed by top management. The information collected by this system is essential for maintaining control of the company.

The foreman's contribution to this system is to insure that all labor and material expended in his group is properly charged and coded.

CHECKING OF COMPLETED WORK

Foremen must check the accuracy, completeness, quality and timeliness of work completed by maintenance craftsmen. The extent of this check varies with the circumstances and depends on the good judgment of the foreman. Figure 3-10 is a guide illustrating accepted practice. There is, of course, no rigid rule for where, when and how often to check work.

It depends on many factors, including:

- Type of work
- Skill of the workers
- Job criticality

TYPE OF WORK / TYPE OF CHECK	NO CHECK	DAILY CHECK	SPOT SAMPLE (AUDIT)	CHECK AFTER COMPLETION	CONTINUOUS SUPERVISION	
MINOR TROUBLE CALLS	X		X			
MAJOR TROUBLE CALLS		X		X		
EMERGENCIES				X	X	
MAJOR REPAIRS		X		X	X	
MINOR CONSTRUCTION				X		
MAJOR CONSTRUCTION		X		X	*	
PREVENTIVE MAINTENANCE	X		X			

* WHEN NECESSARY

Figure 3-10: Checking completed work.

- o Time available for checking

The important thing to remember is that checks must be made to whatever extent necessary to satisfy the foreman that his group is performing with excellence.

CASE STUDY 31: The Linecrew

The example which follows is perfect to summarize this section on productivity control, since it illustrates more of the factors affecting productivity than any true incident I have experienced in my lifetime.

The author was called upon to do a study of the linecrew by the president of a utility company. The function of the crews was to install new transmission lines, new electric service to homes, commercial office buildings and industrial plants and to maintain the electrical distribution system.

I started, as was my practice, by examining the files, work orders, touring the facilities and meeting all of the personnel involved, who gave me the expected promise of their full cooperation and support in my effort. I found very little of value in the files and proceeded to design a form that each foreman was asked to fill out every day, for each job, for two months. In preparing the design of the form, a list was made of all the different types of jobs that the linecrews performed. The various jobs were coded as follows:

S1 = Install service to home
S2 = Install service to industrial plant
S2 = Install service to commercial building
P1 = Install type A pole
P2 = Install type B pole
T1 = Install type A transformer
T2 = Install type B transformer

The complete list coded seventy-five different types of jobs.

The purpose of the form was to record certain information primarily for the period when the crew arrived at the site and started work on the job. A partial list of the information I wanted was:

- o Crew size and trade mix
- o Vehicle number and type
- o Total man hours for each job
- o Material cost

In addition, I intended to ride with the crews during the two month data gathering period to determine what the crew was doing when at and not at the job site. The following is a description of a typical day spent with the linecrews:

Background

There were only two crew sizes, 6 man or 13 man crews, depending on the job size. The smaller crew typically had the following vehicles:

- o Foremen's passenger car
- o Material truck
- o Winch truck
- o Hole digging truck

The larger crew had a proportionately larger number of vehicles. Working hours were 8:00 A.M. to 4:00 P.M., with a half hour allowed for wash-up at 3:30 P.M. and lunch from 11:30 A.M. to 12:00 Noon. The total number of crews was twenty.

Material Acquisition

At the end of each day the foremen made out a list of material needed for the next day. The material trucks were parked side by side each afternoon at quitting time at the supply building loading dock. In the morning all crews arrived at the loading dock; simultaneously, the supply room doors opened and the supply clerks started filling the orders and placing the materials on the loading dock. Each crew would then load its material from the supplies placed on the loading dock.

I discovered to my surprise that the supply building was opened 24 hours/day, 7 days/week. "Why doesn't the supply building night crew load the material trucks?" I asked.

"They don't know how to do it properly," was the answer. It took anywhere from 10 minutes to an hour to load the trucks. Naturally, no truck was allowed to leave until all material trucks were loaded, even if one crew took only 10 minutes to load its truck. This resulted in some crews sitting idle for 50 out of 60 minutes for all the trucks to be loaded.

Coffee Time

At about 9:00 A.M. all crews departed for coffee! There were about 6 coffee shops around town that all men customarily attended. It was a pleasant way to start the day. There was high morale and a great spirit of camaraderie amongst the crews. There were the usual cliques, of course. For example, crews 1, 3, and 7 were very friendly and had coffee in the same favorite spot each day. This closeness was warming to witness. However, on my first day out with the crews, I thought that it was a shame that crews 1 and 3 had to have coffee at the usual spot, 5 miles *east* of home base, when their first job was 15 miles *west* of base! The job location, whether it be north, east, south or west, apparently had no influence on their basic need to have coffee at the same spot each day.

"Where the Heck Is the Place?"

Next, the crew took out the map sketched by the design engineer. Some of the engineers were good at drawing maps; most were not. "Where the heck is the place?" became the standard question. I made a note to talk to the chief engineer about map making.

After the Job--What's Next?

I will analyze what happens on the job after this paragraph. On the first day I was surprised to learn about the following procedure that was practiced after the first job was complete:

- o How long will it take to travel to and complete the next scheduled job?
- o If we tackle the next job, can we get back to home base in time to have half hour clean-up time?
- o If the answer is yes, then the crew proceeds to the next job.

- o If the answer is no, then the crew proceeds back to home base where they wash up and go home.

The result of this standard practice was the crews quit work anywhere between 1:00 P.M. to 4:00 P.M.

The information gathered over the two month period from personal observations and the special forms designed showed that, on the average, the crews' time was spent in thirds as follows:

1/3 Working

1/3 Traveling

1/3 Idle

The Job

The period the crews were actually working was studied by the following technique.

- o Determine before job start, the estimated time duration of the job.
- o Select random times from a table for the period of job performance, i.e., 12:07, 12:13, 12:14, etc.
- o At the random time points, count the number of workers working as opposed to being idle.
- o The numbers working were then added and divided by the number of observations to obtain the average number working.
- o The productivity was then calculated by dividing the average number of men working by the crew size.

The percentage of time working from this productivity measurement showed that the crews had an average productivity of 30%. But, they were only working a third of the time, therefore, the overall productivity was 1/3 x 30% or 10%!

Analysis--General

My job now was to analyze all of the data collected

over the two month period and recommend corrective action. The president was shocked by the 10% productivity and felt certain I was mistaken. It took me about four hours to convince him of the accuracy of my calculations by taking him on a tour to see what was taking place, just as I had done.

Analysis--Job Type Data

Calculations showed that 80% of all work performed by line crews was making service connections.

In watching a service being installed, I concluded it was basically a three man job. One man climbed the pole with his climbing spikes. He worked on installing the transformer and line connection. Another man worked on the connection at the facility. One man was needed on the ground to feed materials to the two men and operate the winch truck to raise the transformer. Based on this analysis, and the fact that service connections amounted to 80% of all work, I recommended that three man crews become the standard size instead of the present standard of 6 or 13 man crews. Larger crews, if needed, could be made by combining two or more three man crews. Furthermore, I recommended the four vehicles presently used by the six man crews be replaced by one bucket truck for all service connection work, thereby saving three vehicles. The three men could easily fit into the bucket truck cab. I also recommended that the foreman become a roving foreman used to spot check several crews instead of having one foreman for each three man crew. The bucket truck could be used not only to lift a man up to the proper working position, but also to raise the transformer. This was a much superior method than using spikes for the men and a winch for the transformer.

The data also revealed that the 13 man crews were occasionally used on service connections which left a minimum of ten men idle.

Furthermore, it was noticed that many jobs were performed by only one man from a 6 or 13 man crew, while the others were idle.

Analysis--Material Acquisition

It was recommended that the night shift stores crew be trained in loading the material trucks and that they load the trucks at night as soon as they were properly trained.

Coffee Time

It was recommended that the practice of stopping at a restaurant be discontinued and, instead, the crew should bring coffee in individual containers or a large single thermos bottle, and the time taken for coffee be periodically audited to insure that only ten minutes be taken in the morning and afternoon instead of the present practice of taking one half to one hour twice per day.

Time to Go Home

It was recommended that the existing outrageous practice of deciding when to leave for home base be discontinued and the crews should leave instead in enough time to allow for half an hour clean-up time.

Maps

It was suggested to the chief engineer that copies of existing sectionalized maps be used to locate jobs and the engineer place a red "X" on the map showing the job location. In addition, of course, he would include the address. The present practice of having engineers draw maps for each job would then be discontinued.

Maintenance

Because of the large influx of people into the city, priority was given to new construction. The result of this was that practically no maintenance work was being done except to handle emergency breakdowns. As a result, the distribution system was in very bad condition and in need of major maintenance repair.

I suggested that the new improvements to productivity should release many men to handle the huge backlog of deferred maintenance work and any layoffs resulting from an increase in productivity be delayed until the backlog was worked off.

SUMMARY OF PART 3

Most managers are aware, knowledgeable, endorse and use the concepts described in Figure 1-1, "Functions of

Management." However, most completely overlook the necessity of using this control system to control the factors of productivity described in Part 3. Prudent managers will not make this mistake and will reap the rewards of increased productivity.

Part 4

Maintenance Support Systems

18

General

In Part 4 selected maintenance support functions will be discussed. A great deal has been written in these areas and therefore the remarks are confined to matters I have learned that are unique, significant or helpful, in my opinion, in managing these support functions with no regard to length or completeness of my remarks. Hopefully you might consider them to be "Words of Wisdom."

Functions covered include engineering, labor relations, training, computers, maintenance materials, spares and work standards. In the case of maintenance materials and spares, it was found that most maintenance organizations do not implement the large number of systems that are available to effectively manage this function. Consequently, these systems are reviewed and listed as a refresher and a few helpful hints are provided for management of the material function. This function is the single factor that most commonly is responsible for low maintenance productivity.

19

Maintenance Materials and Spares

GENERAL

It has been the author's experience in making management studies that material is the single biggest maintenance support function contributing to low maintenance productivity. Why this is true is a puzzle, since material management is a well matured discipline.

Whatever the reason, it is believed that efficient management of the material and spares function is important enough to warrant space in this book. Three of the most common errors committed in material management for maintenance are:

- To have no system of any kind.
- To have no kitting system.
- To not recognize the many different material sub-systems and to leave out management for one or more of these systems. (A list of these systems is given below.)

Three elements are required for the successful completion of a maintenance task: Manpower, Instruction, and Materials--"MIM." This chapter deals with the third element--Materials.

It is important for each foreman to be familiar with all of the systems available for purchasing, stocking, expediting, and obtaining materials in order for him to process his maintenance task with maximum efficiency and labor productivity.

An unavailable one dollar part may easily put four maintenance mechanics out of work for hours while the part is identified, located and delivered. The systems described in this section are designed to minimize this loss. These systems are:

- Receiving
- Kitting
- Main stockrooms
- Sub-stockrooms
- Bench stock--Free issue
- Bench stock--Controlled
- Residual materials
- Surplus equipment
- Scrap
- Reparable tools, parts and equipment
- Expediting
- Inventory management
- Stock catalog
- Spare parts manual
- Material acquisition time

The interrelationship between maintenance foreman and material management personnel for each of these sixteen items is outlined in the following paragraphs.

PROVISIONING

Provisioning is defined as the identification, requisitioning, and stocking of parts that are advantageously stored in the warehouse.

The foreman should recognize there is a cost involved in stocking material over and above the purchase cost. This cost is the return on the money spent for stocking material which would be realized if the money were invested. Because of this the foreman should only request stocking of a part when it is apparent that the savings in equipment downtime and labor productivity indicate an advantage to the company.

With this in mind, foremen should submit provisioning requests to the warehouse under the following circumstances:

- When new equipment is installed, a recommended spare parts list of items to be stocked should be submitted preferably in advance of the start up date.
- When high usage on a currently used outside purchased part indicates that it should be stocked.

When submitting provisioning requests, the foreman should also indicate if the part is to be carried as:

- General stock and spares
- Insurance spares
- Critical dedicated spares

General Stock and Spares

General stock and spare parts are items used generally throughout the plant and are normally of low dollar value and high usage. When provisioning these items, foremen should include a recommended minimum-maximum quantity and an estimated annual usage rate to the chief warehouseman. After the item has been placed in stock, it should be the responsibility of the chief warehouseman to monitor usage, and recommend and process changes to minimum-maximum figures to maintain proper stock levels and to circumvent stock depletions.

Once a year, the chief warehouseman should list items falling below his recommended minimum annual usage rates and decide on corrective action in cooperation with the maintenance and production managers. Possible actions taken will be:

- o Delete item from stock
- o Reduce minimum-maximum levels
- o Reduce minimum annual usage rate

Insurance Spares

Foremen may decide, in the case of production critical equipment, that spare parts should be stocked regardless of the annual usage rate. For example, a set of bearings for a very large motor may be stocked even though the anticipated usage is one set per 4 years.

In this case, the cost of equipment downtime is the overriding factor and takes precedence over inventory expense. These spares are called "insurance spares."

The foreman should designate the quantity to be stocked and the warehouse should reorder on a "one for one" basis. That, is, if one is withdrawn, one is ordered to replace it. No minimum annual usage rate control should be applied to this type of stock.

Critical Dedicated Spares

A particular maintenance foreman may have responsibility for an item of production equipment that is so critical he doesn't want to chance having someone else requisition and possibly deplete the stock on a critical spare for this item of equipment.

He can guard against this possibility by designating the spare as a "critical dedicated spare" on his provisioning request.

Dedicated spares may only be issued to the person originating the provisioning action or his alternate. The spare will be coded accordingly in the stock catalog and a note made of the restriction on the parts bin.

Dedicated spares will receive the same type of inventory control as insurance spares.

If a foreman other than the one authorized to withdraw a particular stock item needs the item, he must convince the authorized foreman to sign a stock withdrawal slip. If the foreman refuses, he may appeal to the maintenance manager.

Responsibility of Chief Warehouseman

The chief warehouseman should be charged with the responsibility of insuring that an "efficient" stock of materials is maintained in the warehouse.

In this case "efficient" means not too high and not too low a level. He shall achieve this by monitoring usage rates.

As a part of this responsibility, he should have the authority to reject any provisioning request particularly of items designated as insurance or dedicated spares, particularly if he feels the use of these categorizations is unwarranted.

It should be emphasized that designations of insurance or dedicated stock should be held to an absolute minimum for the following reasons:

- Insurance stock represents a relaxing of inventory level control and too widespread a use may raise stock levels to an uneconomical point.
- Dedicated spares represent a restriction on stock usage which can hamper operations of people not authorized to withdraw the part.

If a foreman's provisioning request is rejected by the chief warehouseman, he may appeal the decision to the next higher level.

Provisioning Procedure

When a foreman desires to stock a new item he should fill out a provisioning request. See Figure 4-1. This document should include the following items:

- Description of part including manufacturer's part number
- Suggested vendor
- Recommended minimum-maximum
- Stock type designation
- Estimated annual usage

The foreman retains one copy and forwards original to

NO. 0742 PROVISIONING REQUEST DATE________

PLEASE ADD, DELETE, OR MAKE CHANGES TO STOCK AS INDICATED.

ADD ☐ DELETE ☐ CHANGE SPECIFICATION ☐

DESCRIPTION: (Include Mfg's Serial No.)________

SUGGESTED VENDOR (S):________

RECOMMENDED MINIMUM ANNUAL USAGE:________

RECOMMENDED MIN________ MAX________

STOCK TYPE: GENERAL ☐ INSURANCE ☐ DEDICATED ☐

THIS PART IS A SPARE FOR:________

ORIGINATED BY:________ REQUIRED DATE:________

APPROVED BY:________ APPROVED BY:________
SUPERINTENDENT WAREHOUSE

Figure 4-1: Provisioning request.

the maintenance manager for approval. Upon approval, he forwards original to chief warehouseman who:

- o Accepts or rejects request
- o Assigns stock number and location
- o Fills out stock card and warehouse stock purchase requisition

The same form should be used to process recommendations for deleting parts from stock.

MAINTENANCE OF STOCK LEVELS

Once an item has been approved for stocking, it should be the responsibility of the warehouse to maintain adequate stock levels.

This should be accomplished by observing stock levels while making entries on the stock cards. When a stock level is too low, the warehouse should pull the warehouse stock purchase requisition and send it to purchasing for procurement action. Normally the maximum quantity should be ordered; however, if the warehouse has first hand knowledge of an unusual one-time requirement considerably in excess of the maximum storage quantity, he may up the minimum reorder point by this quantity to circumvent long term stock depletion.

PURCHASING

General

There are four main items of concern relating to satisfactory purchasing service to the maintenance function. These are:

- o Adequate communications between maintenance and purchasing departments on the criticality of purchase requisitions (PR's).
- o Adequate system of reporting on status of PR's.
- o Adequate statistical information for use in measuring the purchasing department's performance.
- o A reasonable delivery time on regular and repetitive stock purchases.

The purchasing department is a *SERVICE* organization. The success of a service organization cannot be measured only in terms of its internal operating efficiency, but must also be measured by its impact on the efficiency of the customers it serves. The latter is by far the most important. This fact is sometimes overlooked by purchasing agents attempting to reduce labor costs of a few purchasing people at the expense of decreased efficiency of thousands of people they serve. The purchasing agent's goal should be the same as that described for maintenance in Figure 1-2, Chapter 4. Refer to case study 1-D. Comments and suggestions related to the above four items of concern follow.

Communications

One of the common faults that I have found in the purchase requisition (PR) document is that there is no place for indicating the importance of the task requiring the material nor the required delivery date.

It is suggested that a box be provided on the PR for both the priority and the required date. The priority marking should be identical to that used on Work Orders.

Once the PR has been assigned to a buyer, he should assume *total* responsibility for becoming familiar with the PR criticality and for obtaining the material on schedule.

Frequently the responsibility for meeting material delivery dates is diluted amongst several people in the purchasing department.

It is further suggested that a box be set aside on the PR and purchase order (PO) for the work order in order to facilitate identification of items in the warehouse when received for kitting purposes. The PO should specify that the vendor mark his shipping document with the work order number.

Difficulty is experienced in receiving when the vendor ships a different item than that specified on the PO. This occurs under the following circumstances:

- o When an "or equal" term is used on the PO.
- o When vendor changes part nomenclature.

This difficulty in identifying parts in receiving can be eliminated if the PO gives instructions to the vendor to state that item "XYZ" shipped is equal to item "ABC" ordered.

PR Status

Every requisitioner of material has the right to call up purchasing and inquire into the status of his order.

Frequently practices in purchasing make it difficult for them to give an expeditious reply. The reason it is difficult may be because:

- They do not log PR's as they come in and therefore are not sure they have received it.
- If they have received it, it is difficult to find because it may be in several places or with several people or in several files. These must all be searched.
- The several pieces of paper involved in processing one PR are filed in different files.

The lack of systemization in keeping track of the processing of a PR causes the following problems:

- Makes it difficult to report status to requestor.
- Makes it difficult for purchasing themselves to follow up on PR processing.
- Makes it more cumbersome for the responsible buyer to assume total responsibility for expediting the PR assigned to him.
- Makes it very difficult to accumulate statistical information for the purpose of measuring purchasing department performance.

In order to solve this problem, it is suggested that a log be maintained. A recommended log is shown below as Figure 4-2:

1	CONTROL NO.	0001	0002	0003	0004	0005	0006	0007	0008	0009
2	DATE ASSIGNED	5/1								
3	PR/STOCK NOS.	4268								
4	QUOTE NO.	1234								
5	QUOTE DATE	5/4								
6	QUOTE RCVD DAY	5/18								
7	PO NO.	3456								
8	PO DATE	5/23								
9	EST COST PR	9999								
10	PRIORITY	4								
11	PR REQUEST DAY	6/1								
12	DATE RECEIVED	6/5								

Figure 4-2: Purchasing department–log book.

This log was designed to achieve the following:

- o PR status reporting capability
- o Internal processing control
- o Performance measurement capability
- o Least amount of paperwork

Comments on this log follow. (Any log that serves the same purpose will do):

Item 1, Control Number

The unit transaction in the purchasing department shall be the control number. It shall be a five digit sequential number starting with 00001 on January 1 of each year.

A file folder shall be made out for each control number with the control number marked on the index tab. *All* paperwork for one control number shall be filed in this folder, including correspondence, original requisitions, quotations, and purchase orders. In other words, each transaction shall be identified by *one* folder, located in *one* file cabinet, and *one* buyer shall be assigned to the responsibility of purchasing the materials for this transaction on time.

When a transaction is complete, it shall be transferred to a complete, single filing system, identical to the in-process file. This file shall be identified by year of transaction. All of the above is designed to circumvent the frequent practice of splitting the responsibility for the PR amongst several people and also splitting up the many documents related to one PR into several places with several people in charge of the several documents. When this happens, it is impossible to pin-point responsibility.

When an original PR is received in purchasing a control number shall be assigned from the log book. In order to notify the requester to use this number when making inquiries on status, a pre-printed notification slip shall be utilized as shown in Figure 4-3 below. The requestor shall staple this form to the PR. He shall fill out his return address and his requisition number. When the control number is assigned it shall be entered on this slip and mailed back to the requestor. This slip achieves the following purposes:

- o Enables use of a sequential control number.
- o Informs requester of receipt in purchasing of his request and of its control number for future inquiry purposes.

When repetitive purchase requisitions are received, they are sorted and grouped by like vendor. Control numbers and folders should be made out for each group. It is recommended that the control number identification slips *not* be used for repetitive requisitions. However, the stock numbers covered by each control number should be entered in the log in column 3 in place of the PR number. When any person removes any paperwork from a file folder, he should insert a pre-printed identification slip in its place in the folder indicating who has the missing material.

Item 9

It is suggested that an estimated total cost be included

INTER OFFICE MAIL

NAME: H. Hendrickson
TITLE: Plant Engineer
ADDRESS: Electrical Shop

FRONT OF FORM

Your original requisition has been received . Please use the indicated control number for all inquiries regarding this transaction.

John Smith
CHIEF BUYER

24689	00001
ORIGINAL REQUEST NUMBER	CONTROL NUMBER

BACK OF FORM

Figure 4-3: Control number identification slip.

on the original requisition. This will serve the following purposes:

- o Enable person signing the requisition to have a rough idea of the dollar value of what he is approving.
- o Provide a quick means for purchasing to determine if bids should be obtained.
- o To facilitate statistical reporting of purchasing activity by dollar value groupings.

Provide communication between the requester and purchasing on the criticality of the order.

The above logging and processing procedures has the following advantages:

- o Provides status reporting capability
- o Reduces clerical effort
- o Provides statistical information for performance measurement purposes

Purchasing Performance

The log book outlined in the previous paragraph will provide most of the information needed for measurement of purchasing department performance. The following paragraphs indicate potential use of the log book information.

Comparison of Date Required and Date Received

Suggest monthly plot of percentage of transactions completed on time as an overall barometer of performance.

Quotation Time

Suggest a periodic audit of this time (item 6-5). This information can be used to reduce time in some areas as opposed to a fixed policy of allowing 2 weeks.

Item 2-8

This information can be used to determine detailed and generalized total paper flow time in purchasing for corrective action purposes.

Item 3

May be used for determining ratio of repetitive stock transactions versus other type transactions.

The processing problems of each type differ. This ratio could influence the type of processing system. If the ratio indicates a predominance of numbers of one type, consideration may be given to designing the system to facilitate this type. If the ratio indicates equal distribution, a single compromise system or two separate systems may be indicated.

Item 10

If use of emergency procurement action is considered to be abused, this column will indicate the magnitude of the abuse and the trend of corrective action. Conversely, comparison of this item with columns 11 and 12 will enable measurement of performance to emergencies. Finally, this column will enable comparison of percentages of transactions by priority and delivery performance by priority.

Item 9, Estimated Cost

The column will enable periodic audit of numbers and percentages of transactions by cost ranges, i.e.:

COST RANGE IN $	*PERCENTAGE OF TOTAL*
0-1	XX
1-5	XX
5-10	XX
10-20	XX
20-100	XX
100-500	XX
500-1000	XX
1000-5000	XX
Over 5000	XX
	100%

Many things can be done with these figures. For example, if the average cost of processing a PR plus a PO was $100, then you would not want to use this system for purchasing a $1 part. Instead you might suggest use of a petty cash fund. Similarly, other cost ranges might suggest other procedures.

Item 1

Allows measurement of number of daily transactions.

These are a few ways that the log book can be used for measuring purchasing department performance and trends and for precipitating corrective action.

Repetitive Purchase Orders

In an effort to reduce purchasing costs, many companies resort to the use of annual or repetitive purchase orders. This advantage must not be offset by faulty procedures or flow times which are illustrated by the following comments that I made related to a purchasing activity that I was consulting on:

Statistics indicate that the average delivery time on repetitive stock purchases is 72 days.

On positive minimums, it forces minimum values that may be higher than desirable in order to circumvent stock outages.

Several other problems aggravate the problem. These are:

- o Annual purchase order percentage is less than 1%.
- o No expediting preference is given to repetitive purchases for stock items that are out of stock.
- o Stock minimums are partially determined by average delivery times in order to circumvent stock outages. There is no evidence that performance to these delivery times is monitored.
- o Flow time of repetitive orders is increased by three procedures that should be studied in detail with a view towards total flow time decrease.
    - Batching of orders in the warehouse (orders are sent to purchasing on the average of every 3 to 5 days resulting in increased flow time).
    - Batching of orders in purchasing to consolidate orders by vendor resulting in increased flow time.

- Certain orders are sent out to various department heads for determination of order. quantity. The advantages of this procedure should be weighed against the disadvantage of increased flow time and its resulting impact on maintenance performance.

The main point inferred in the previous comments is that the processing time of repetitive orders is not only a major problem affecting maintenance performance but that it frequently is an unmeasured problem.

Accordingly, three main action items are suggested:

- That a flow time measurement program be implemented.
- That a detailed flow time study be initiated.
- That the percentage of annual purchase orders be increased. This will have the following advantages:
  - Purchasing department labor hours will be reduced in direct proportion.
  - Promise of an annual order to a vendor should be a wedge tending to decrease delivery time and therefore decrease stock outages with resulting increase of material availability to maintenance.
  - Total flow time will be decreased.

WAREHOUSING

General

The warehouse function is responsible for the orderly receiving, inspecting, storing and issuing of maintenance and other materials. The systems used in carrying out this responsibility are outlined in the following sections. The systems are designed with that old proverb in mind, "A place for everything and everything in its place." Only by making optimum use of all of these systems can we achieve a neat and orderly looking plant while maintaining the most economical use of labor as far as it is effected by the material function.

Receiving

The warehouse receives parts from the following sources:

- Outside purchases made for non-stock materials
- Outside purchases made for stock materials
- Residual parts returned
- Reparable equipment, tools, and parts
- Surplus equipment
- Scrap

Figure 4-4 illustrates the flow of materials into and out of the receiving area.

Incoming scrap should be unloaded into a special holding area for this purpose in the scrap yard.

Incoming surplus equipment should be unloaded into a special holding area for this purpose.

Outside purchases, reparables (repaired parts) and residuals (surplus materials) should be delivered to the main receiving area in the warehouse which should be separated into marked off areas A through E as shown in Figure 4-5.

The importance of having these marked off and segregated areas cannot be overemphasized. They create not only a visual impression of efficiency but tend to impart to the warehouse workmen a pride in their area, a sense of

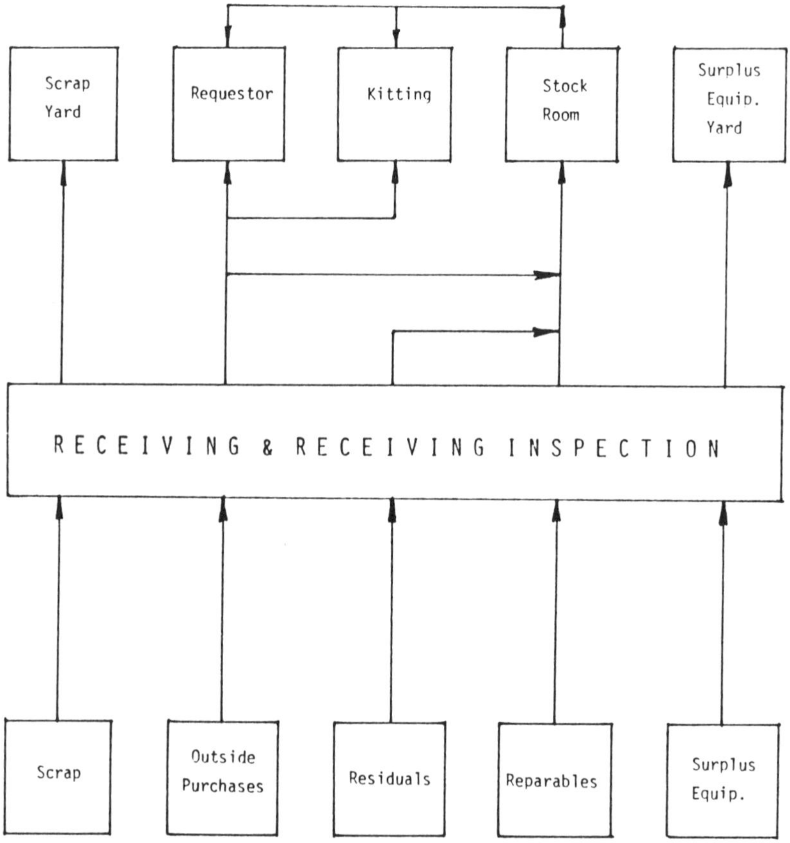

Figure 4-4: Receiving and receiving inspection.

order, a visual flow path, and consequently an actual increase in efficiency.

Kitting

Kitting is defined as the process of obtaining all materials for a specific job in advance of the job start date.

A full "kitted" work order enables the job to proceed without interruption due to material unavailability.

Construction, modification, rearrangement, and major maintenance work (Priority 3 work) can be conveniently kitted in advance due to the fact that this type of work can normally be planned well in advance of its need date. Because of this and the beneficial effect kitting has on labor productivity:

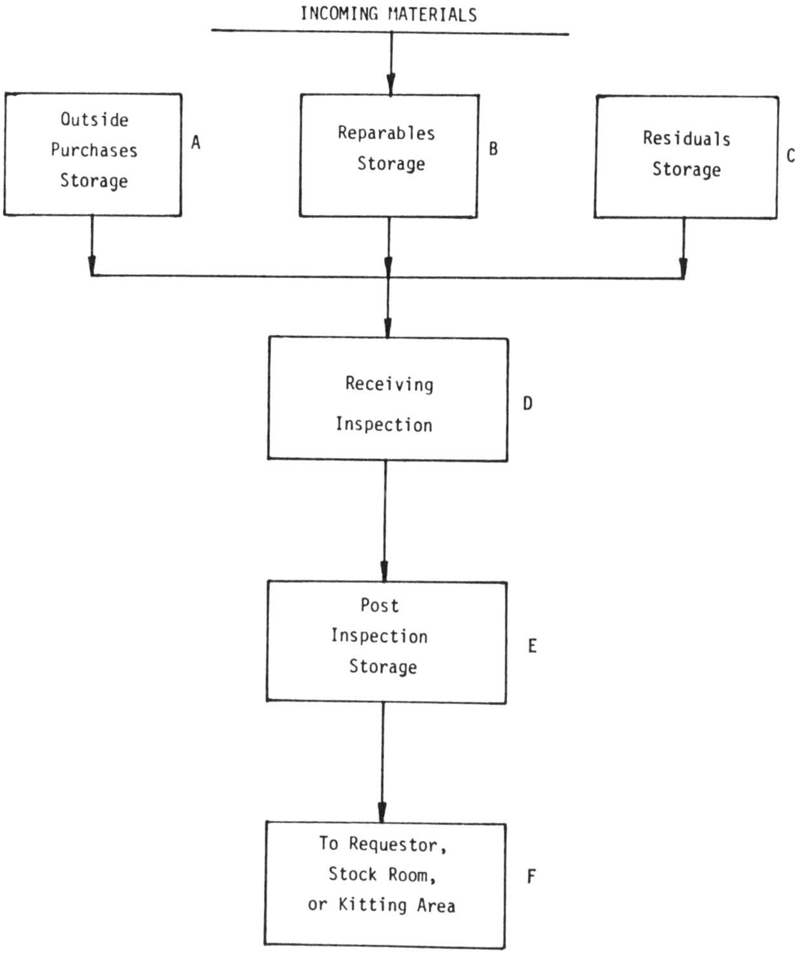

Figure 4-5: Receiving area flow.

ALL PRIORITY 3 WORK SHOULD BE KITTED IN ADVANCE

Priority 1, 2 and 4 work may be kitted at the discretion of the maintenance foreman.

"Kitting" should be accomplished by utilizing the following procedure:

Foreman's Responsibility

o After work order has been prepared and

approved, the foreman or engineer should prepare a bill of materials (see Figure 4-6).

- o Prepare purchase requisition for all outside purchases. (A note should be placed on the purchase requisition to have the purchased item tagged for the work order.)
- o Send bill of materials to warehouse.

BILL OF MATERIALS

DATE__________ WORK ORDER NO.__________

ITEM NO.	QTY.		STOCK NO.	PURCHASE ORDER NO.	DATE RECEIVED

SHEET_____ OF_____

Figure 4-6: Bill of materials.

Warehouse Responsibility

- o Establish manila file folder for each work order which is to be kitted.
- o Insert copies of bill of materials, purchase requisitions and *purchase orders* in folder.

- o Write stock requisitions and fill all stock items on bill of materials.
- o Designate kitting area for work order and mark location on work order kitting folder.
- o Deliver stock items to designated kitting area.
- o Receive, check off, and deliver outside purchases to kitting area.
- o When entire bill of materials is checked off as complete, call foreman and notify him to pick up completed kit.

Foreman's Responsibility

When the work order is complete, the foreman should notify the warehouse who should close out and discard the work order kit file.

Main Stockrooms

Functions normally performed at the stockroom are:

- o Receiving and receiving inspection
- o Kitting
- o Stocking and issuing of materials and parts
- o Stock part requisitioning
- o Inventory management

Sub-Stockrooms

Under certain ,circumstances wisdom indicates that advantages can be realized by creation of a sub-stockroom at a remote location. When a sub-stockroom is used, the parts stocked may be:

- o Parts held exclusively in the sub-stockroom
- o Duplicate stock normally with different minimum-maximum levels

The sub-stockroom may be managed by warehouse personnel in a similar manner to the system used for the main warehouses.

An alternate method of providing a sub-stockroom is as follows:

- o Locked access to the sub-stockroom will be controlled by the maintenance foreman.
- o Issues will be made by the maintenance foreman. The requestor will fill out a stock requisition in an identical manner used at the main warehouse.
- o Stock requisition slips will be forwarded on a daily basis to the chief warehouseman who will post to stock cards located at the main warehouse office.
- o Reordering, restocking, binning and locating of stock in the sub-stockroom will be the responsibility of the warehouse.

Bench Stock--Free Issue

Certain materials are used so frequently and are of such a low cost per unit that it would be uneconomical to purchase or requisition the item from stock one at a time.

An example of this type of item would be nuts, bolts and washers.

The system for handling these items is called *free issue bench stock.* It functions as follows:

- o The foreman decides what items logically fall within the category of free issue bench stock.
- o The foreman provides adequate bins and shelves for storage.
- o The foreman requisitions the parts from stock on a periodic basis and charges parts to the store's account.
- o The mechanic helps himself to as many bench stock parts as he needs on a free issue basis with no paperwork.

Foremen must be cautious in their decision to identify a part as free issue bench stock and recognize that this action represents a loss in material control and a temptation to personnel who may have use for the part at home. The advantages and disadvantages of this system must be kept in balance and periodic audits conducted of usage rates.

Bench Stock--Controlled

Where the bench system is advantageous but the part is either valuable or most particularly adaptable for private use, the "controlled" bench stock concept may be used. This system is the same as free issue bench stock, except the part is kept locked up either in a cabinet or in a locked and fenced in area. As in the case of free issue bench stock, there is no paperwork transaction, however, the person needing the part must obtain the key from an authorized foreman.

Residual Materials

Residual materials are those purchased for a particular task in excess of quantities needed either because of overestimating or because of changed job conditions.

When residual materials are discovered by a foreman, he may use one of the following procedures:

- Put item in bench stock bin
- Scrap item
- Return item to stock if it is in units of issue

Surplus Equipment

Surplus equipment should be defined as assembled, integral units of equipment that are surplus to current operating needs and which are presently not in use. The equipment may have come from one of the following sources:

- Change in operating mode
- Over order on construction job
- Ordered as a spare

The warehouse should be responsible for managing the surplus equipment yard. This yard should be divided into the following areas:

- o Unresolved area
- o For sale area
- o For re-use area

All equipment identified as "surplus" should be tagged with a red tag, identified as "surplus equipment--unresolved," and delivered to the unresolved area of the surplus equipment yard where the equipment should be numbered and logged in by the warehouse attendant.

A disposition of what to do with the unit, i.e., re-use, sell or scrap, should be made by the appropriate area maintenance supervisor.

Surplus equipment to be sold should be tagged with an appropriate yellow tag and located in a separate area.

The warehouse should provide an indexed listing of all surplus equipment available for re-use.

Scrap

The warehouse should be responsible for managing the orderly flow of scrap materials from generation, collection and storage, to disposition.

The collection and storage system should be designed to segregate scrap into logical categories to facilitate sales, for example:

- o Ferrous
- o Stainless
- o Copper
- o Aluminum
- o Rubber
- o Wire
- o Lumber
- o Rope
- o Machinery

Reparable Tools, Parts and Equipment

When a maintenance foreman is faced with the task of

putting a broken down item of equipment into operating condition, he has two choices:

- o Repair the damaged item
- o Renew the damaged item

When the latter choice is made, he can do three things with the damaged part:

- o Scrap it
- o Rebuild it and use it as bench stock
- o Rebuild it and return it to stock if it is a stock item

The last case is the preferred course of action since it will minimize the inventory level and provide for a neater and cleaner plant.

STOCK CATALOG

Introduction

Most stockrooms in small plants have practically no thought given to organization and in particular have no stock catalog. Maintenance of a Sears Roebuck type of catalog can reap benefits in reduced time for maintenance men to get materials particularly when it is managed by a good stockroom man who not only knows what he has, but where it is and what its name is. He will also know what parts he has for a particular machine.

How many times has a mechanic come to the stockroom looking for a part, but doesn't know its name, or knows its name and can't retrieve it because the part is listed under a different name? A good stockroom man will know *all* the names a part is called by, just as the parts man knows in an auto supply store. But you will notice that this man always has a complete and accurate stock catalog at his disposal.

The pages that follow describe a typical stock catalog and spare parts manual that I designed for a client that can be used for any plant with suitable modifications.

General

A stock catalog shall be maintained for all materials and parts stocked in the warehouse.

The following items of information shall be identified in the catalog for each item stocked in the warehouse.

- o Stock number
- o Item description
- o Item location
- o Stock type

Description of a Stock Number

The stock number shall be an eight digit number with logic as follows:

```
1234 -- 567 -- 8
XXXX -- XXX -- X
 |       |     |___________ STOCK TYPE
 |       |
 |       |_________________ PART NUMBER
 |
 |_________________________ CLASSIFICATION NUMBER
```

Alpha (letter) and numeric (number) characters shall be used for each digit as required. Alpha characters shall be A to Z omitting the "O," and numeric shall be 0-9.

Digit 1: Digit 1 shall be used to define major stock classifications:

Code Number	Description
1	General
2	Electrical
3	Heavy equipment
4	Machinery
5	Drill machinery
6	Locomotives

Digit 2: Digit 2 shall indicate sub-major classifications as shown on Figure 4-7 below (2nd digit in table):

Digit 3: Digit 3 shall indicate the first letter of the alpha description of the stock item in alphabetical sequence A to Z. For example, digit 3 for stock item "bolts" shall be coded "B."

CODE	DESCRIPTION
10	Assay supplies and chemicals
11	Auto and truck parts
12	Bearings, pillow blocks etc.
13	Belts, belting and supplies
14	Bolts, nuts and washers
15	Castings and liners
16	Cement, sand and gravel
17	Chain, rope and fittings
18	Drill steel and drill bits
19	Explosives
1a	Hardware
1b	Hose and fittings
1c	Instruments, and meters
1d	Oil, fuel and greases
1e	Packings
1f	Paint and supplies
1g	Pipe and fittings
1h	Rail and fittings
1i	Reagents
1j	Refractories
1k	Steel and misc metals
1l	Timber
1m	Tools, hand

CODE	DESCRIPTION
1n	Valves and parts
1p	Welding Supplies
20	Cable, electrical
21	Electrical supplies
22	Electrical parts
23	Electronic supplies
24	Hoists
25	Locomotives, battery
26	Locomotives, trolley
27	Trolley supplies
30	Heavy equipment
40	Balls and rods
41	Ball and rod mills
42	Boilers and burners
43	Cars, ore transportation
44	Compressors, air
45	Crushers
46	Fork lifts
47	Hoists
48	Loaders
49	Machinery
4a	Pumps
4b	Saws
4c	Tools, pneumatic
50	Drill machinery
60	Locomotives

Figure 4-7: Chart of stores accounts.

Digit 4: Digit 4 shall be coded 0-9 and A through Z for breakdowns of digit 3. For example:

B-0 bolts, square head

B-1 bolts, stove

B-2 bolts, slotted

Digit 5, 6, and 7: Digits 5, 6, and 7 shall be used for specifying the detailed part number.

Digit 8: Digit 8 shall be used for identifying the stock type as shown in Figure 4-8 below:

Catalog Arrangement and Index

The catalog shall be arranged and indexed by the stock number in order of each digit of the number sequentially.

The index shall be cut off on digit number 4 and page number references inserted after each classification.

CODE	DESCRIPTION	REORDER POINT	USAGE CONTROL	USE DESCRIPT.
1	GENERAL STOCK AND SPARES	MIN-MAX CONTROL	ESTABLISH MIN YEARLY USAGE	NONE
2	INSURANCE SPARES	REORDER ONE FOR ONE	NONE	NONE
3	DEDICATED SPARES	MIN-MAX CONTROL	ESTABLISH MIN YEARLY USAGE	CAN ONLY BE WITHDRAWN BY MAN AUTHORIZED

Figure 4-8: Stock classifications.

Stock Location Number

The stock location number shall be a seven digit number arranged as follows:

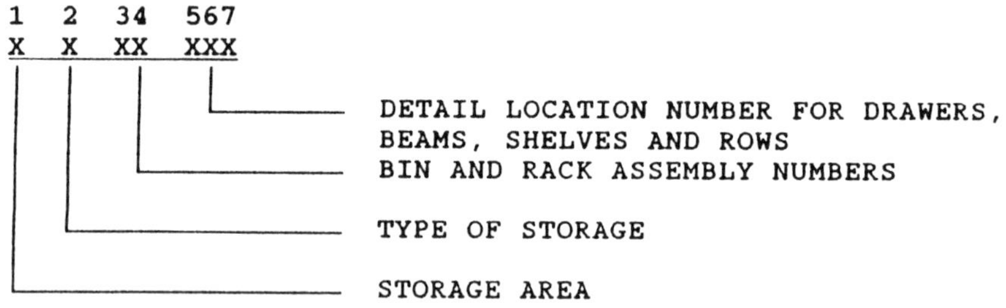

Computer Input Sheet

The chief warehouseman with the help of maintenance foreman shall transfer information from each stock card to the computer input sheet shown in Figure 4-9 below:

STOCK NUMBER	LOCATION	DESCRIPTION
21F00111	PB16761	FUSE, 10A, TYPE 1AZ

Figure 4-9: Computer input sheet.

This operation will require designing of a stock number and location number for each stock item in accordance with the procedures described in this section. Careful consideration should be given to the logic and conciseness of the word description. Once the three items have been fixed, they should be recorded on the computer sheet *and* the stock card.

As the computer sheets are completed they should be forwarded to the computer department for entry.

Computer Program

The computer department shall design a computer program to print a catalog indexed and arranged as described previously.

Printing, Error Correction and Updating

The catalog shall be printed on a quarterly basis. Errors discovered by warehousemen and foremen should be recorded and given to the chief warehouseman for processing through the computer section. Stock additions and deletions should be handled in a similar manner. These changes to the catalog will be processed on a quarterly basis.

SPARE PARTS MANUAL

General

A spare parts manual should be maintained to help maintenance foremen in the expeditious identification and location of spare parts.

Each maintenance foreman has the responsibility for provisioning the quantity of spare parts necessary for the economical maintenance and minimum down time on equipment that falls under his maintenance responsibility.

This section describes the method by which the manual is originated, updated and printed.

Index Number

In order to provide for computerized printing and update capability, a six digit index number shall be used as follows:

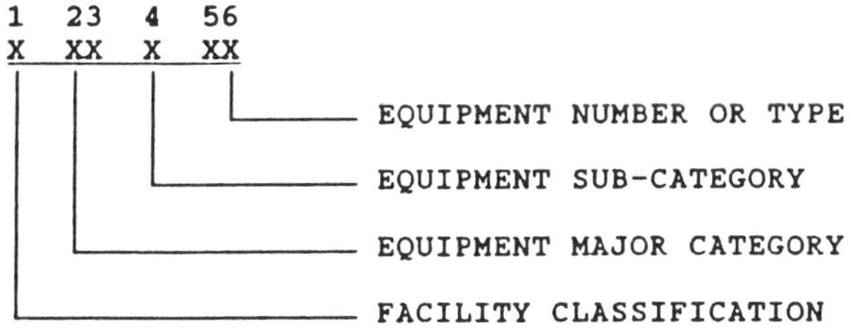

Single digit classifications shall start with zero and proceed to nine; the next sequence will be A proceeding to Z but excluding the letter O to eliminate confusion with the number 0.

Two digit classifications shall proceed with the first digit fixed with the second digit varying from 0-9, then A to Z again excluding the letter O. After Z is reached the first digit will be raised to 1 and the cycle then repeated until the first digit reaches Z. The first number of a two digit series shall be 00; the last ZZ. This method will provide for 1,225 separate identifications with 2 digits.

Digit 1: Digit 1 shall categorize facility classifications as shown in Figure 4-10 below:

The "general" category, code 9, shall list all spares not included in the other categories, for example, hospital and administrative building parts.

Concerning electrical spare parts, they shall be listed in two primary places as follows:

By Equipment

Mechanical, electrical and instrument spare parts shall all be listed under the item of equipment they service. For example, crusher parts shall be listed under *Mill*, then under *Crusher*.

Electrical Distribution System

All other spare parts shall be listed under code E, "Electrical Distribution System." This listing is primarily for systems that are 100% electrical.

Interface

An interface diagram showing the break point between the two categories of electrical described above is shown on the following page as Figure 4-11.

CODE NO.	DESCRIPTION
0	MINE
1	MINE
2	MINE
3	MILL
4	SMELTER
5	REFINERY
9	GENERAL
B	HEAVY EQUIPMENT
C	RAILROAD
D	POWER PLANT
E	ELECTRICAL DISTRIBUTION SYSTEM

Figure 4-10: Facility classifications.

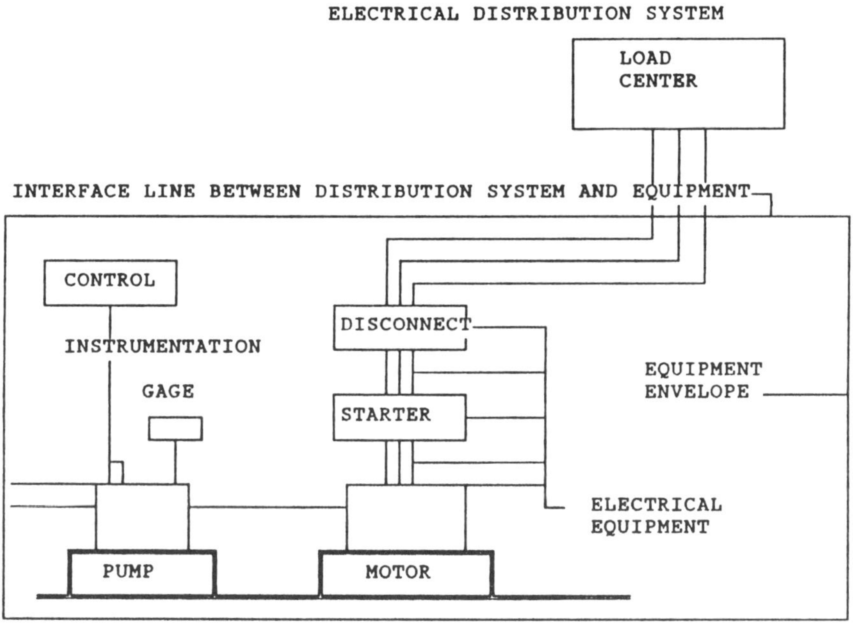

Figure 4-11: Typical electrical interface.

Digits 2 and 3: Digits 2 and 3 shall describe major equipment categories as shown in Figure 4-12.

REFINERY

DIGIT 2 & 3	EQUIPMENT MAJOR CATEGORY
0 0	AGITATORS
0 1	BANDING
0 2	BURNER
0 3	CASTING BOWL
0 4	CHARGERS
0 5	CLASSIFIERS
0 6	CONVEYORS
0 7	CRANES
0 8	DRYERS
0 9	ELEVATORS
0 A	FEEDERS
0 B	HVAC

Figure 4-12: Major categories of equipment.

Digit 4: Digit 4 shall divide digits 2 and 3 into sub-categories as shown in Figure 4-13.

REFINERY
HVAC CODE OB

DIGIT 4	EQUIPMENT SUB-CATEGORY
0	BOILER, STEAM
1	BOILER, WATER
2	FAN, EXHAUST
3	FAN, ROOF
4	FAN, GENERAL
5	HEATER, SPACE

Figure 4-13: Equipment sub-categories.

Digits 5 and 6: Digits 5 and 6 shall be used to identify a *specific* individual item of equipment. If there are several identical items of equipment this fact should be surfaced in the "description" part of the code identification. See Figure 4-14.

Equipment Number Digits 5 & 6	Description
00	Babcock and Wilcox boiler model 1A8Z (parts for boilers nos. 01 and 02 are identical
19	Fan, exhaust, Sturtevant Model 20
AX	Space Heater, Trane Model 240

Figure 4-14: Equipment number.

Computer Input Sheet

After the maintenance foremen have categorized and identified all items of equipment that require spare parts, as outlined in the previous paragraphs, they shall draw up a list of spare parts required for each of these equipment items. This list shall be made on a computer input sheet for this purpose shown in Figure 4-15 below:

EQUIP. NO.	STOCK NO.	LOCATION	SPARE PARTS LIST
30B000			
	NONE		AMBER RING NO 10104
	NONE		BREECHING CLIPS MODEL A
	48510002	PB1701	SAFETY VALVE, CRANE 204
	48T10003	PB1702	TUBES 1"
	48T20013	PB1702	TUBES 2"
	48T30023	PB1703	TUBES 3"

Figure 4-15: Computer input sheet.

After the list is made up, the corresponding stock number shall be inserted as indicated. If the part is not stocked, the word "none" shall be recorded as indicated. If the foreman desires to have the part stocked, he should initiate provisioning action.

Manual Arrangement and Index

The manual shall be arranged and indexed by the equipment number in order of each digit of the number sequentially.

The index shall be cut off at digit no. 4, and page number references inserted after each classification.

Computer Program

The computer department shall design a computer program to print a manual indexed and arranged as described above.

Printing, Error Correction and Updating

The manual shall be printed on a quarterly basis. Errors discovered by foremen should be corrected by means of the computer input sheet. Updating should be handled in a similar manner.

20

Engineering

In many plants the man in charge of maintenance is the plant engineering manager who is also in charge of facilities engineering in addition to maintenance. Engineering is an important support service to maintenance since it frequently is used to analyze failures particularly when a redesign is necessary and to write preventive maintenance instructions. Both maintenance and facilities engineering are service organizations and therefore the plant engineering manager must please his clients (all the other managers in the plant) if he wants to be successful. For some reason many plant engineering managers recognize that a consulting engineer must please his clients, but never realize that they must too. Some have a tendency to be dictatorial in dealings with their clients.

CASE STUDY 4A: The Newcomer

I had been at work for six months in my new job as plant engineering manager of a large industrial plant with an employment of 50,000 people. I had a staff off 1,000 people of which 150 were engineers. It was my first big job. I was only 30 years old, a nervous wreck and starting to crack under the strain. I was the first new manager hired in 20 years. All the other managers were 55 years of age and older. I had a feeling they resented the young "newcomer." To increase my concern, the general superintendent of production, with whom I was friendly, told me that I was in trouble with the old timers. He said all of them wanted a couple of engineers from my central engineering group to be assigned to them permanently to work under their direction so that they didn't have to go through my backlog control to get their jobs done. I explained why I was resisting these requests. I had the younger, less experienced engineers working as "short order" engineers (a term I coined) which meant engineers working as specialists on small, recurring jobs. An example of this work was wiring up new machinery and running new compressed air outlets. We handled thousands of these. As a result of the highly repetitive work these engineers soon memorized the designs and cost estimates so that they could do the work in a fraction of the time that it would take the more experienced engineers I reserved for the larger more complex jobs. I argued that if I acceded to all the requests, there would be nothing left of central engineering and that the decentralized effort with 1-3 people scattered through the plant would require them to know all engineering disciplines, as opposed to the specialist concept I used. This "jack-of-all-trades" concept would mean they would take longer to do a job and the engineering stood a greater chance of being botched. Furthermore, since there would be peaks and valleys to their workload, they might frequently be out of work because of the lack of a large backlog to work to. The superintendent said he was impressed with my argument but warned me that the old timers would probably do me in if I didn't bend a little. I was worried.

Concerning decentralization versus centralization, I have a theory about that. I believe that companies decentralize when they are making money because this form of organization achieves tighter control of activities and is more responsive to the needs of the group receiving the support. However, it costs more because it is more expensive for reasons stated above. When companies are losing money, they centralize because this type organization is more cost effective again for reasons stated above.

Two days later the president of the firm dropped in to see me. He said, "You're in trouble with the old timers." My face turned white. The phone rang. It was someone yelling at me for not having his machine back in service. I stood up and paced back and forth with the phone cord dragging for five minutes as we argued. The president later said he noticed I *always* pace back and forth when I talked on the phone, and it was important for me to project an image of calmness in the face of crisis. I tensed up even more.

He then lectured me for one hour on the need to sell myself to the "old timers." He suggested I visit every manager in the plant once per month with notebook in hand and say, "I would like to know what I can do to improve my service to your department." When I started doing this, they were all stunned! "No other plant engineering manager has ever been interested in my problems!" they all said. I wrote down what they said and did my best to handle their problems. I even gave away some engineers to the managers who were most insistent. In about four months, I was accepted by all and my problems were over.

This was perhaps the most valuable lesson I ever learned and I am grateful to the president of the firm for having taught it to me. The phone incident taught me the need to exhibit a calm exterior at all times, regardless of pressure, a trait which I diligently strove to perfect.

The plant engineering manager is always in the middle concerning job priorities. No matter what logic he uses, someone will always accuse him of playing favorites. "Who do you think you are deciding that *his* job comes before mine?" is the standard complaint. One way out of this, if you can get all your clients to agree, is to set priorities by committee. You start off by coming up with a weekly strawman priority list. Then you hold a meeting of all the managers and let them critique and finalize the list. After this is done, then they can't yell at you because it is *their* list. The problem is that they probably will not want to spend the time to do it. Nevertheless, you should try it in writing. If they turn it down, then you can say that the alternate is for them to accept your list without rancor.

CASE STUDY 4B: Thirty More?

At a plant where I was Director of Engineering, Operations and Maintenance, over 3,000 people, the work load was increasing fast. I already had 100 people in the planning and scheduling department and put a request in to management for 30 more. "Thirty more?" they said, "You must be out of your mind!" I explained that the number of jobs handled per month had risen dramatically. "No way will we give you more planners and schedulers," they said.

The scheduling operation was chaotic. Management insisted that *all* jobs be formally planned and scheduled two weeks in advance; the first week being firm and the second tentative. This included time phasing the various trades on projects.

Management wanted us to be able to give our customers instant status on all work. This was a big job in itself. It soon became apparent that every week almost 100% of the schedule was altered because of changes in priorities, absences and other reasons. The entire system seemed like a big waste of time.

I spent the next week tabulating work load data. This showed that 90% of the number of jobs required only 10% of the number of planning and scheduling manhours and 10% of the number of jobs required 90% of the manhours. Furthermore, the majority of the small jobs required less than 8 hours to plan. I saw the solution by using the "Magic Tool" concept and made this proposal to management which was accepted:

Small Jobs: On small jobs (90% of the number of jobs), I would set aside 10% of the planners, schedulers and craftsman to handle these jobs. Since the planning averaged less than 8 hours, all of these jobs would be planned in one day with zero backlog. The physical work would be accomplished within two weeks and would not be formally scheduled. Furthermore, no status information would be maintained or given out. The customers would be told simply that his job would be finished in two weeks or less.

Large Jobs: These jobs would be handled with the normal scheduling system complete with formal scheduling, craft phasing and status reporting.

Critical Jobs: These jobs demanded zero schedule slippage and therefore had to be treated with the highest degree of sophistication. For example, if a control valve needed to be manufactured and delivered by a certain date, the scheduler had to track it through production from the beginning so that he could take action if the production schedule slipped. If necessary to meet the valve delivery milestone date, the scheduler might have to pirate a valve or ask an engineer to redesign it in order to get one on time.

The change in the handling of small jobs allowed us to *decrease* the staffing by 25 instead of having to hire 30 more people! This was the second most important lesson I learned concerning the managing of an engineering (and scheduling) organization. This same principle applies to many other functions.

It can be summarized as follows:

> When a major portion of the number of tasks performed are small, requiring a small percentage of the total staffing, then set aside a dedicated crew to handle the input each day with zero backlog. This makes your clients happy and relieves you of the scheduling and status reporting workload and allows you to concentrate on the few larger jobs (the "Magic Tool" concept).

How To Be A Success

I would like to give you my philosophy on how to be a success in any job. The philosophy is that there are two things that you must do to be a success in a job, these are:

- o Do a good job at performing the work described in your job description.
- o Appear to be doing a good job at performing the work described in your job description.

The first skill is a job skill and is the one most often thought of when people think of what it is that you are supposed to do.

The second is a political skill, and is a different skill altogether but is equally important.

Some people are good at one, some the other. To be

really successful you must be good at both. Some people are outstanding at politics and not at doing their job and make out better than the person who is performing his job with excellence.

I recall one incident related to this from my experience. I felt that one of my smoothest and efficient running operations was my maintenance storeroom and receiving department. However, my boss chose to chastise me for running a sloppy department. I thereupon set upon a program to achieve a visual impression of efficiency which I admit it did not have. I did the following things:

- Painted the bins in different colors by function, i.e., kitting, spares, stores, reparables, residuals, and so forth. Each type of bin had a different color and each type had a sign designating what type it was.
- I placed a sign outside the receiving dock that read, "UNLOAD INCOMING SUPPLIES HERE."
- I painted lines on the floor showing the flow of incoming materials through receiving and receiving inspection.

When I was all through, my boss complimented me on running an outstanding department. Nothing had changed in the operation except that it now had a visual impression of efficiency.

I'm sure that a book could be written on how to achieve these political skills but I have written this to emphasize the importance of the skill to those who might otherwise overlook it. In general it relates to getting your boss to think you are doing a good job regardless of what kind of job you are actually doing. For example, a very introverted person who never advertises his accomplishments coupled with a boss who doesn't go out of his way to find out what the person is doing may result in the impression that the person is dong nothing. This is unfortunate but is a fact of life. Salesmanship and political astuteness are necessary to be a success!

21

Labor Relations

Americans have been brought up with the concept that there is a natural conflict between the company, represented by management, and labor, represented by the blue collar workers. Each is out to exploit the other.

To a lesser extent, private industry looks upon the government as the enemy, an organization bent upon overtaxing, overregulating and installing restrictive organizations such as EPA, OSHA and the like, so that private industry can hardly stay in business.

Well, look out, private citizens! In countries such as Japan the opposite situation exists! Management and labor work together for each other's mutual benefit and the government passes laws and provides aide to private industry to help it be successful. At the end of World War II, the U.S. was the leading shipbuilding, steel making, car producing and electronics nation in the world. This rank has been lost to Japan. Other industries have fallen to Japan and still others are about to.

Maintenance foreman must do their share in inviting cooperation between management and labor to the point that it is obvious to both parties that they share a common interest, the *long term* success of the company. Maintaining control over labor relations problem is a function of all supervision even though the overall guidance rests with the director of labor relations. As far as the maintenance foremen are concerned, there are two important points to remember:

- *The Labor Relations Triangle:* There should be a mutual feeling of trust, confidence, and respect built up between the foreman, the worker and the union steward.

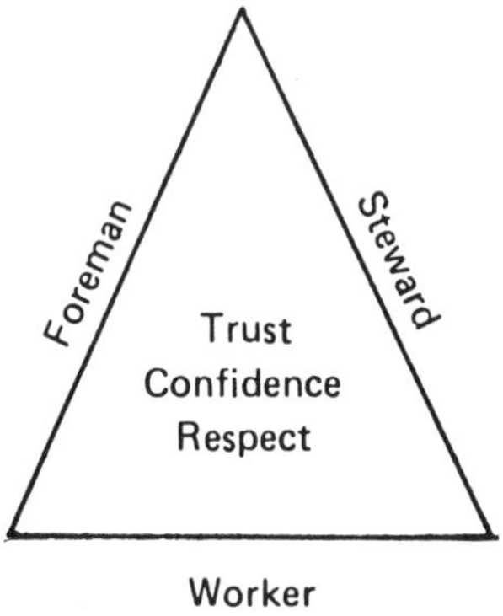

The worker should feel that he can go to his foreman with his problems, and that the foreman will listen to him and give him help when possible. If the foreman doesn't do this, the man will turn to his steward exclusively. Under these circumstances, the foreman has lost control of the labor situation and opened Pandora's box.

- *Contract Negotiations:* Contract negotiations are not the prime concern of maintenance foremen. Nevertheless, it must be remembered that the unions always have a long range goal (plan) towards which they are actively working. These plans will generally include higher wages, more fringe benefits, greater job security and more jobs. The last two items are frequently achieved by:

    - Increased specialization and more job categories

- o Decreased flexibility in inter-plant transfers
- o New-hire restrictions

These inevitably mean reduced productivity and therefore greater employment. The company must have a long range plan to counter these unfavorable trends and to convince the union that their mutual goal is the long term success of the company, *and* a pleasant, rewarding working environment for the worker. This is the job of the labor relations department. Nevertheless, they can use help from maintenance foremen who have more direct experience with the operational difficulties caused by restrictive contract clauses and who may have ideas for trade-offs or other suggestions for improvements to the contract. The labor relations department needs this free flow of ideas, suggestions and gripes to be in a strong posture during negotiations. To achieve this level of planning and communication is to be in control of the labor situation.

In addition, the company must recognize that hourly workers are the closest to the operation and its problems. They are therefore in an excellent position to see the problem areas and to recommend improvements. Furthermore, their productivity is a major factor affecting economic success. Their productivity is *heavily* influenced by their morale and attitude toward the company, their boss, their job, working conditions, and motivational programs. These must always be enhanced by a cooperative relationship between management and labor as opposed to the attitude illustrated in Case Study 3C.

In summary, it is critical for us to note that the conflict relationships throughout American businesses that maintain economic and social distance between producers and managers and among firms, unions, government, consumers, owners and financial institutions are virtually absent in Japan. The contrast is startling.

The principal lesson for the United States is that it can no longer tolerate the higher costs of its adversary relationships. In the final analysis, achieving high national productivity may depend more on efficient, harmonious organization than on efficient technology. We are all in this international competition together, and the most socially efficient nations are very likely to be the winners.

22

Training

GENERAL

In Chapter 2 it was mentioned that the first priority in any maintenance organization is to have a crew of mechanics who can repair anything in the plant that breaks down. This need and priority is obvious. The function that insures this capability, as the plant is modified and updated, is training.

There is an abundance of material available on the subject of training, i.e., textbooks, plant engineering training manuals, apprenticeship programs, supervisory training, etc. Two unique contributions are offered which are the direct result of observations made in the field of discrepancies in the training program.

At Kennedy Space Center as we were about to start the countdown on the first launch to the moon, I realized we were about to perform certain operational and maintenance tasks that were never performed before. Certain of these tasks were simple whereas others required specialized training. The complex tasks were reduced to written procedures and the men repeated the tasks over and over again until they performed them perfectly. I wanted to make certain this skill was recorded for use by backup personnel and therefore, developed the technique described in this chapter. It is the author's belief that all plants have a need for this technique for certain specialized tasks.

Another problem noticed in the field while on assignment as a consultant were the large number of plants having complex machines, just installed, without a suitable training program for the mechanic assigned to maintain it.

This gets to be a big problem if the plant has several dozen different type machines without a mechanic who understands how to maintain them. Frequently this problem is the *only* training problem! The company may have an outstanding apprentice, supervisory and other training programs and lack only the training in specific complex equipment. A system for training for this type is also described in this chapter.

TRAINING IN SPECIAL SKILLS

Usually there are dozens of special maintenance tasks in a plant that are not only critical to production but also difficult to perform from a manual dexterity and technical skill aspect. These types of tasks can be taught with the aid of sound movies or video tape.

The technique is similar to that used in the medical profession when they take movies of a surgeon performing some new and difficult operation. Every aspiring surgeon says "If I could only see an expert perform one of these operations, I would feel less hesitant about performing one myself for the first time." The problem may be the particular operation he wants to see is performed rarely and that there are very few experts in the world to perform it. This certainly was the case with the first heart transplant. The same situation arises in the maintenance field.

The following procedure illustrates how the technique is used. Let us imagine that the special task is the dismantling, overhaul and reinstallation of a gland seal on a turbine. Assume that the seal is very complex with different types of metal, fine finishes, carbon rings, close tolerances, etc. Let us assume that the total task takes 2 hours.

Step 1

Obtain the most qualified mechanic in the plant in this particular task.

Step 2

Obtain a qualified cameraman complete with a Super 8 sound movie camera or video camera. This could be a foreman or mechanic.

Step 3

Take sound movies of the task. Instruct the mechanic to describe what he is doing as he is performing the task, giving special warnings, precautions, and how-to hints as he is going along.

This movie would then be shown to other mechanics who would absorb the expertise of the mechanic making the film.

The wonderful part about this training method is that it is so magnificently inexpensive.

In the case shown above, the only expense is the two hours' time for the cameraman and the film cost. The mechanic's time isn't counted since he is working on an actual job.

It goes without saying that this technique should be limited to specific complex critical tasks where training films are not already available from outside sources. The maintenance manager should decide where to use this technique.

TRAINING IN COMPLEX CRITICAL EQUIPMENT

Let us conjure up a special training problem. Assume that the XYZ Paper Cup Company has only one kind of machine in the entire company, a paper cup making machine. Two types of craft training programs are being considered for the maintenance worker:

1. A one week training program in maintenance of the paper cup making machine. (A discussion with the manufacturer revealed that this is all the time they need for a complete training course.)

2. A general correspondence school course in mechanical maintenance coupled with on-the-job training lasting three years.

Clearly program 1 will produce results faster and cheaper in this case.

Almost every plant has cases of equipment that justify specific training in the equipment on hand, in preference to or in conjunction with generalized training.

Specific training of this type should be restricted to equipment items having all of the following characteristics:

- o Equipment is critical to production
- o Equipment is complex technically
- o Breakdown repair records show a high failure and downtime rate

The entire point of this paragraph is to emphasize that in some plants, the heart of the maintenance problem is maintenance know-how on *very few* equipment items and, when this is the case, specific training is indicated rather than generalized training.

The specific training course should be developed by the plant expert in this equipment and the manufacturer and should contain the following:

- o A description of thc equipment and its controls.
- o A discussion of its criticality and downtime cost.
- o Preventive maintenance tasks including overhaul instructions.
- o Trouble analysis in this format:

  Trouble?

  Probable cause?

  Remedy?

- o Operating instructions including how to operate in various component failure mode situations.

It is unfortunately a fact that many companies are too prone to dump new equipment on mechanics without bothering to determine if they know how to repair it and without giving them any training with the mistaken belief that competent mechanics can repair anything.

RELATIONSHIP BETWEEN OPERATING AND MAINTENANCE SKILLS

Operational and maintenance know-how of equipment is inter-related. A maintenance man cannot be successful without having a knowledge of equipment operation. An operating man must have some knowledge of maintenance problems on the equipment he is operating, at least to the extent that operation affects maintenance problems. Operational abuse resulting from lack of knowledge of the machine can cause damage and downtime.

Figure 4-16 illustrates the learning process that a maintenance man must go through in order to have a thorough understanding of the equipment he is responsible for maintaining. Comments on this diagram follow.

The first step in the learning process is to learn the names of all equipment components. This step (1A) automatically gives an insight to its purpose (1B), how to operate it (1C), and a small idea of how to maintain it (1D).

The next step is to learn the purpose of each component. This step (2A), automatically gives further insight into operation (2B), and maintenance (2C).

The next step is to learn how to operate it. This step (3A) does not provide complete operational knowledge since it does not provide knowledge of the maintenance impact on operating techniques. However, this training does provide further knowledge of maintenance (3B). The operator must gain the minimum knowledge of maintenance (4B) required to prevent operator abuse.

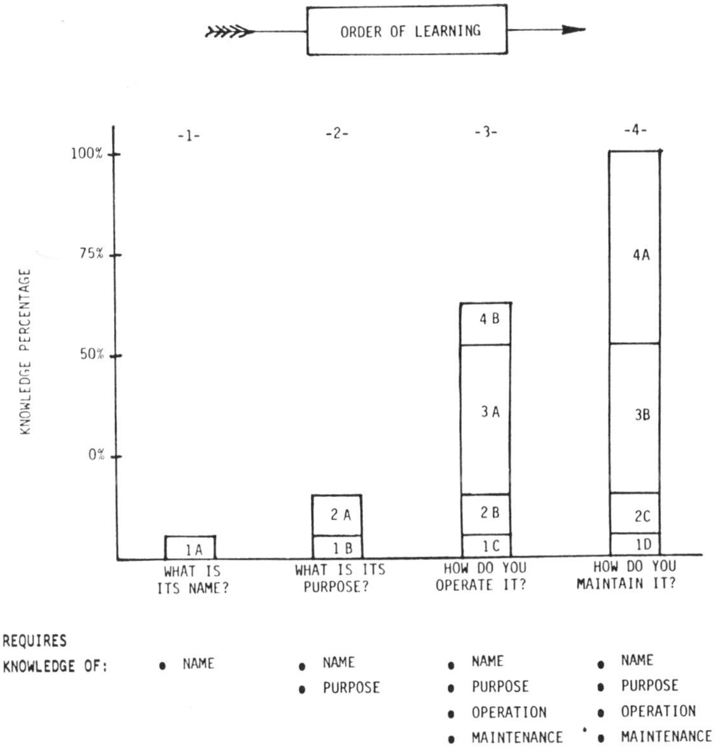

Figure 4-16: Operation and maintenance learning process.

The final step (4A) is to learn how to maintain the equipment. This completes the maintenance know-how requirement. It is desirable for maintenance personnel to have 100% knowledge of all four of the discussed items. Operational men should have 100% knowledge of the first three and sufficient knowledge of maintenance to be aware of the operational effect on maintenance.

Many equipment failures have pyramided into excessive downtime primarily as a result of an operator's lack of knowledge of the correct operational response when equipment suffers a complete or partial breakdown. It is quite obvious that knowledge is required in the case of an airplane, a ship or a nuclear plant, but the need for it is sometimes overlooked in a plant.

Case study 3D illustrates the problems associated with inadequate training of operators.

23

Computer Applications

In my many years of experience as a maintenance management consultant and plant engineering and maintenance manager, I have had many opportunities to observe computerized maintenance management systems. For the most part, these systems were inoperative or ineffective. Many of the systems were being modified for the second, third, or fourth times and still were not working properly. Frequently the system supplier had been working on the system for years without success. I found that there are common reasons for these failures:

1. *THE DESIGNER OF THE COMPUTER SYSTEM HAD NO EXPERIENCE MANAGING A MAINTENANCE DEPARTMENT*

This is the kiss of death to a system, whether or not the designer is from the plant staff or an outside vendor. When using the plant staff, the error most frequently committed is to assign the design to a computer expert or someone from the finance and accounting department with little or no participation from the maintenance department. When designed by the finance department, the system normally is oriented to accounting, not to maintenance. On outside purchased systems, the vendor may have no actual experience managing a maintenance department. (Why not ask the vendor for a resume of the man who designed the system?) If the maintenance manager is not involved in the design process because the system is being forced on him, or because he feels defensive about his lack of computer expertise, the result is a system that will "provide you with every conceivable piece of information you could possibly want." The result of this is that a 2 to 4 foot thick tab run comes out weekly that no one could possibly read and doesn't. Everyone is afraid to say that the system is worthless, and therefore the system is perpetuated for years causing great inefficiencies, wasted dollars, and worse, diversion from more important work. In conclusion, it is essential that a computer system be designed by someone with extensive direct experience in managing maintenance, who also has computer expertise.

2. *THE WORSHIP OF FALSE IDOLS*

Some people feel that the following systems will automatically bring about great improvements in efficiency, quality and morale:

- o Computerized systems
- o Complex budget control systems
- o Work standards

These false beliefs, practiced in a manner similar to unreasoning idol worship, have led many astray. One of the reasons I wrote this textbook was to dispel these beliefs. My book emphasizes the philosophy that good management is the application of three common sense systems which are:

- o Maintenance optimization
- o Downtime improvement
- o Productivity control

Only when the fundamentals of these systems are in place and working properly should more sophisticated systems be considered. Computerized systems should be installed as a useful aid to simplify (not complicate) a functioning sound common sense system. If the maintenance system is ineffective due to poor design or inadequate management, computerizing it will probably only make it worse and increase costs.

Most computerized systems are designed for very large complex multi-plant applications in the belief that only very large operations can afford or benefit from computerization.

When these systems are used in small plants, no one bothers to remove the superfluous components with the result that personnel are turned off by the inappropriate complexity and the system is not used. I HAVE FOUND THAT A NEED EXISTS TO DESIGN AN EASY TO USE AND PRACTICAL SYSTEM TAILORED TO SMALL AND MEDIUM-SIZED PLANTS and designed in such a way that anyone can operate it without fear of being inundated with paperwork.

3. *MATERIAL SYSTEM INCLUDED WITH MAINTENANCE PROGRAM*

Many maintenance computer programs include all the steps necessary to process materials for each work order. This can include inventory management, min-max control and procurement, for example. I strongly advise against this practice, particulary for the small and medium-sized plants. The reason is that material management is so complex that it detracts from careful consideration of the maintenance work order which is complex enough by itself. Instead I recommend that just a total material estimate and actual cost be put on the work order to allow the approver to have a feel for the total cost. If you do install such a system with material management as part of it, mark my words, you will live to regret it.

4. *TIMEKEEPING INCLUDED WITH MAINTENANCE PROGRAM*

Some maintenance programs also include timekeeping. This also can lead to disaster with endless and fruitless checking of the time reported on time cards against time reported on work orders. They never will match and if you employ this system, you will be wasting many hours on fruitless effort.

My philosophy concerning estimating labor and material and the amount of time you spend on it is plainly stated in Chapter 24. I am unsympathetic with spending too much time at it since a very accurate estimate in itself *accomplishes nothing* particularly if the problem is that the mechanic does not have the things he needs to get the job done properly. I have found this to be the case in all of the plants that were in enough trouble to need my consulting services.

As I have stated in the preface to the second edition, people who have read the first edition of my book have strongly encouraged me to design a computer program that utilizes the systems in this book. This I have done and the system is now available from my company. Appendix A describes the system and how to use it. A free brochure and three demo disks at nominal charge can be obtained by writing to: *Frank Herbaty and Associates, 253 Plainview Drive, Bolingbrook, Illinois 60439.*

24

Maintenance Work Standards

GENERAL

"Maintenance Work Standards" is a method of estimating maintenance work using basic industrial engineering principles just as they are applied in production operations. In principle, work is measured with standard time data by breaking down a job into its basic components and applying time standards for each component. The sum of these standard times, plus various allowances for job conditions, represents the time that should be allowed for the job. A comparison of this allowed time with actual time measures the performance or job effectiveness.

"Maintenance Work Standards" or "Universal Maintenance Standards" have the potential of being a very useful tool to the maintenance manager. They also can cause a waste of time and money and a loss in productivity if used improperly, since some people may forget that production operations are highly repetitive while maintenance operations are not.

"THE WORST CASE"

The following example illustrates "The Worst Case" of the use of work standards in maintenance and assumes the standards are applied to all work and every job. This is an actual case from my experience as a consultant.

First, the maintenance crew performs the maintenance task and reports the hours spent on the job. This is the "actual time" spent. After the job is complete, the standards applicators are dispatched to the job to calculate how much time the job should have taken according to their set of work standards, which can be a set of books varying in thickness from 1 to 12 inches. This time is the "standard time" spent. The amount of detail used in determining the standard time is very great. For example, if a mechanic takes apart an eight bolt pipe flange, the standards book will show how many minutes or seconds it should take to remove one bolt from the flange under various conditions, i.e., when the flange is new, in use for five years or when severely rusted. The standards applicator must then add up these hundreds or thousands of time increments to arrive at the total standard time. The standards also provide times for obtaining materials, tools, equipment, travel time, idle time, union business, personal time and many other factors. Standards applicators are frequently industrial engineers with little or no maintenance expertise, and in this case, must first get the maintenance men to show them the job site and explain how the job was performed. This is further complicated by the fact that many of the jobs include opening up a piece of machinery, fixing it, then closing it up. In this case, the standards applicator can't *see* what was actually done, nor the problems encountered inside the machine when it was opened up. Furthermore, the greatest majority of maintenance jobs are of very short duration; because of this and the procedure described above, *it may take more time to estimate the standard time than it actually takes to do the job!* This is particularly serious if the procedure is followed for *all* maintenance work, which is sometimes the case.

The actual time is divided by the standard time to give a productivity factor. This information is frequently averaged for each foreman to show the relative ranking of each foreman and to promote a spirit of competitiveness between them so each foreman will strive to obtain the highest productivity factor. This procedure may cost you hundreds of thousands of dollars to implement and millions to operate together with a loss in overall productivity if you don't understand how to properly use and apply work standards.

In this plant where I was called in to effect improvements and to recommend steps to achieve a 20% reduction in maintenance personnel because of a drop in sales, I had *no* problem with my first recommendation. The recommendation was to eliminate all 35 people working on the standards program. I explained that not only would this result in the immediate savings associated with the 35 man layoff, but that further savings would be achieved by the increase in productivity I forecasted resulting from the cessation of the disruptive activities of the standards group who applied standards to everything instead of using them only on the few cases where they made sense.

Remembering that estimating maintenance work is an essential element of an effective maintenance program, and recognizing that engineered maintenance work standards is one of the estimating tools, when and how should we use this tool? The following paragraphs answer these questions.

AUDITING PRIORITIES

When a plant is suffering from low productivity, it is important that the problem be investigated in the priority order described in Chapter 2, Audit Priorities.

The reason for this is that the problem is more likely to be the total absence or ineffective implementation of one or more of the fundamental management systems, rather than poor estimating practices.

WORK SAMPLING

Work sampling is a very simple tool to use to determine where time is being spent by categories, i.e, travel, material and tool acquisition, idle, working, etc. A work sampling study should be performed at least once per year for each foreman to show if there is an improving or deteriorating trend. For a plant not well organized, this study will show that most of the time spent is non-productive and only a small percentage of time is spent working. Work standards are normally not used to perform the work sampling study. After the study is complete, each of the factors affecting productivity should be carefully analyzed (see Chapter 14) to determine which factor is the biggest contributor to low productivity. This factor should be improved first and then each other sequentially until time spent working is high enough to justify a study of the work method.

An alternate to work sampling is 100% sampling or time study of an activity. An example of this type of analysis is illustrated in case study 3I, "The Line Crew." Whenever possible, I prefer to use the 100% sampling method since you see everything related to the job activity and are therefore in a much stronger position to determine problem areas and the solutions to the problem.

Some people might be intimidated by the expression 100% and feel that it takes too long and is too expensive. The amount of time spent is usually not a problem for two reasons:

- Use the method only if you sense serious problems.
- Use the method only on activities where a one-time study is sufficient.

ESTIMATING ROUTINE REPETITIVE WORK

Normally, routine repetitive work need be estimated only once, not every time the job is performed. A perfect example is janitorial work. When a new building is about to be put into service, the estimator makes a list of everything that needs to be cleaned, i.e., water closets, lavatories, floors, walls, windows, etc. He then uses his janitorial work standards book to estimate the manpower required. There are many of these standards available. Janitorial work represents the best and most useful application of maintenance work standards. After the manpower is selected, then its up to the foreman to train his men in the work and audit their performance periodically.

Another method of estimating routine work can be explained by using the preventive maintenance daily inspection route sheet work as an example. (The same method can be used for janitorial work.) The method proceeds in steps, i.e.:

Step 1

Make a rough estimate using your best judgment.

Step 2

Assign manpower to the task and train them in the work until you are satisfied that they are performing the work properly.

Step 3

Pick the man doing the best job and simply time him to see how long he takes working at a reasonable pace. This time then becomes the estimate and "standard" time.

A final method of estimating may be used on highly repetitive jobs such as installing electrical outlets or compressed air lines. The method starts with recording the actual times on work orders (no estimate is made at all). After a period of time, say, six months to a year, the amount of time for all work orders on the highly repetitive task is totalled and then divided by the number of work orders. This is the average time spent and the number to be used in the future for estimating the task. Since these type of jobs are normally of very short duration, and the variables associated with the job high, this method is good enough for scheduling and control purposes. The planner scheduler should do this for all highly repetitive tasks. People who use this method sometimes dictate a small improvement percentage to the standard of perhaps 1/4 of 1% over a period of 6 months or a year.

ESTIMATING CONSTRUCTION AND MODIFICATION WORK

The estimating of construction and modification work is a highly developed skill requiring great expertise in many engineering and construction disciplines. There are many reference books, such as *Means Site Work Cost Data*, which are available and give figures for both labor and material. These books are updated periodically.

It is essential that the estimator incorporate, and know how to apply percentages for escalation and contingency, particularly for large dollar value jobs of long duration.

Since construction estimating is such a complex and critical skill which is seldom found in the typical plant engineering and maintenance department, it usually pays to hire outside specialists to do the estimate on the very costly jobs, and at times, use them to check the estimate of the architect/engineering firm or the company design staff. Life cycle costing and value engineering techniques should be used to provide a sound basis for approval of the construction job. Modification tasks employ the same techniques, however, demolition is involved to a much greater extent. In the case of demolition, the estimator must anticipate and study tasks that are "hidden in the walls" lest he leave out a high cost item. Decontamination of hazardous radioactive or toxic wastes are particularly susceptible to underestimating.

ESTIMATING MAJOR MAINTENANCE JOBS

Most major maintenance jobs are basically non-repetitive. Each one is either completely different or at least slightly different than the last job. Parts of the job are known and can be estimated and other parts are unknown and cannot be determined until the machine is opened for inspection as explained in Chapter 15. Major jobs are estimated utilizing a combination of construction estimating aides, past experience on similar or identical jobs and if available, engineered work standards. The most important of these is past experience. There is no easy way to gain this experience other than by hard work, training from experts, the gathering of estimating data, and the passage of time spent working on the estimating task.

ESTIMATING REARRANGEMENT WORK

There are no books on estimating rearrangement work! The man skilled at estimating rearrangement work relies on talents developed in estimating the types of maintenance work previously described and primarily in experience gained in doing this type of estimating for several years and comparing estimated versus actual costs on completed jobs.

ESTIMATING TROUBLE CALLS

The foreman has two options concerning estimating of trouble calls. One is not to estimate them, since the scope of the job is unknown. The other option is to write "guesstimate" on the trouble call slip. The latter option is recommended. It provides the foreman with a rough estimate of whether or not he is giving the mechanic a day's worth of work and approximately how long he will be gone on the job. If he estimates the job will take one hour and it actually takes sixteen hours, he is entitled to an explanation. Furthermore, the mechanic should know that someone cares whether or not he spends sixteen hours on a one hour job. If he finds out no one is interested, and if he is not a conscientious worker, you may find him sleeping in some secluded spot. If the mechanic's explanation is reasonable, then you can forget it. If not, you may have to discipline or instruct him. In either case, the mechanic should record the actual hours expended on the trouble call slip so that records may be kept of the total hours spent on this type of work for estimating purposes and since the percentage of time spent on breakdowns is a measure of the success or failure of the preventive maintenance program.

SUMMARY

In summary, let me repeat the main points to be remembered when considering the use of engineered maintenance work standards:

- When productivity is low, it is normally because of the complete absence, or the inadequate application of the several management functions described in this book, not because of poor estimating techniques. In other words, give the men the tools, equipment, materials, instructions and outages they need to do the job and arrange to get them to the job expeditiously before you start worrying about how long it takes them to do the job! Study the factors affecting productivity (Chapter 14).

o Estimating *is* important! But one must recognize that the estimating system varies with the type of work and that engineered work standards is just one type of an estimating system. Each estimating system must pay for itself in improved productivity. Engineered work standards is one of the more sophisticated systems and, therefore, should be used with knowledge and understanding of its contribution to productivity.

Part 5

Fleet Management

25

Introduction

In many ways fleet equipment can be managed in the same manner as any other plant equipment item. However, the differences are great enough to warrant a separate section in this book. Some of the major differences are listed below:

- Equipment moves throughout or outside the plant.
- PM is normally scheduled on the basis of hours operated or miles driven rather than elapsed time.
- The prime mover is normally an internal combustion engine rather than an electric motor and therefore exhausts pollutants and must carry its own fuel.
- Some equipment may travel all over the United States and even to foreign countries.
- Equipment is normally depreciated over a much shorter time period.
- Safety can be a bigger problem because equipment is moving.

Throughout this section on fleet management, references will be made back to sections previously covered that apply equally well to fleet management. To start off with the entire Part 1 on Maintenance Optimization applies. Most of the philosophizing on why things are done has already been done in the first four parts of this book; therefore this section on fleet management will emphasize *how* to do things for fleet equipment.

To further simplify the description of the various fleet management processes, this text will present the material as if it were part of a manual on fleet management. The author has conducted many management studies concerning fleet operations. All of these studies have surfaced several common unsatisfactory areas which are:

- Unsatisfactory and unused computerized management information systems.
- No PM system.
- No written policy and procedures.
- No historical records.
- No failure analysis system.
- Mechanics working on the wrong shift.
- No material system.
- Poor vehicle utilization and no system for insuring proper utilization.
- Excessive Make/Type combinations.
- No scheduling system.
- No training program.
- No work order system.
- Inadequate system for justifying, acquiring or disposing of vehicles.
- No management control system.

The use of the systems, forms and procedures for Fleet Management described in the following chapters should eliminate all of the above unsatisfactory areas if properly installed and implemented.

26

Maintenance

GENERAL

It is important to keep in mind the fact that maintenance costs are a relatively small portion of the cost of owning and operating a vehicle. This point is illustrated in the following table which is a reprint of a study on passenger vehicles conducted by AAA. This shows maintenance costs at 4.71%.

COST OF OWNING AND OPERATING A PASSENGER VEHICLE

COST ITEM	PERCENT
TIRES	3.61
MAINTENANCE	4.71
GAS AND OIL	19.10
DEPRECIATION	47.61
INSURANCE, REGISTRATION AND LICENSE	24.97
TOTAL	100.00

Although maintenance costs are a minor percentage, they have *the* major impact on production lost time and vehicle downtime! In other words, maintenance has the biggest impact on the carrying out of the mission of the vehicle.

Therefore it can be seen that percentagewise, the difference in cost between a very poor maintenance program and a good maintenance program is relatively small, considering the overall cost.

In summary, there is no justification for a poorly-managed maintenance program. A well designed preventive maintenance program will:

- o Reduce vehicle downtime
- o Reduce maintenance costs
- o Reduce production lost time costs

CATEGORIZATION OF EQUIPMENT BY CRITICALITY

All equipment contained on the fleet equipment inventory list should be categorized by criticality as follows:

Category A - Critical Equipment

This includes equipment such as fork lift trucks, the lack of which will cause a serious impairment to operational capability. Another characteristic of the category is the lack of spare units to turn to in the even of vehicle breakdown. Consideration may be given to including vehicles which are important from a "political" or "public relations" viewpoint. The PM program for these units should be designed to achieve close to zero breakdowns. In other words, more emphasis would be placed on achieving low downtime rather than minimum cost because operational lost time is a much larger cost than maintenance cost in this category.

Category B - Normal Equipment

This would include the majority of equipment which has an important function, but is not considered to be critical. Vehicles that have spares and those which are susceptible to pooling would fall into this category.

A normal PM program would be designed for this category of equipment with the purpose of preserving and protecting the vehicles and insuring a reasonable downtime percentage at minimum maintenance expense.

Category C - Non-Critical Equipment

This category would include those vehicles which would not seriously impair the plant's function if down for repairs. This equipment would be given last priority in the PM program. If the total labor capacity is tied up on the PM program for categories A and B, the PM program for category C vehicles would be dropped or performed by an outside garage.

The criticality code designator, A, B, or C, shall be included on the inventory list as shown in the paragraph on this subject.

Coding shall be the responsibility of the garage superintendent with the aid of the various department heads. Since the coding effort is so important and will result in continuously preferential treatment to certain vehicles, the final coding shall be approved by the Plant Engineering Manager.

Compliance with the scheduled preventive maintenance work on Category A equipment shall be mandatory. A special study shall be made of PM parts requirements on this class of vehicles to insure preplanned kitting and availability of materials needed for PM work prior to the scheduled PM work date.

If there is insufficient labor capacity to perform work other than the mandatory class A PM work, then one or more of the following alternative methods shall be used:

- Drop PM work on class C equipment.
- Send work to outside shops.
- Work overtime.

The garage superintendent shall consider each situation as it arises and make a judgment of which alternate is most appropriate at the particular time.

CODING OF MAINTENANCE WORK TYPE

All maintenance work should be coded according to the maintenance work type as follows:

Work Type Account Code	Work Type Description
1	Preventive Maintenance
2	Breakdown Maintenance
3	Operation

Preventive maintenance work is only that work performed as a result of the preventive maintenance order. Breakdown maintenance shall include all other maintenance work except operation work described in the next paragraph.

Operation shall include labor and materials used in service station type work, such as:

- Pumping gas
- Fuel, oil and grease, labor and material costs
- Washing and steam cleaning
- Road calls for other than breakdown repair purposes
- Car pickup

Unanticipated tire repairs such as flats should be charged to breakdown repairs. Preplanned tire rotation or replacement specified on PM instructions should be charged to PM.

Repairs made as a result of a PM inspection when the item repaired is not mentioned in the PM instruction should be charged to breakdown maintenance and listed on the breakdown maintenance history card.

EQUIPMENT INVENTORY

General

A list of vehicular equipment should be maintained for the purpose of identifying the type and assignment of vehicles. The garage superintendent should be responsible for the inventory maintenance and distribution of the list to appropriate personnel. The following recommendations for the equipment list are taken from a study that I did for a municipality and is used because it is illustrative of the variety of equipment contained in some fleets. The process used for maintaining and updating the inventory list should be computerized. (See Chapter 23).

The list should contain the following information:

Item No.	Description
1	Equipment Number
2	Equipment Description
3	Acquisition Cost
4	Make
5	Model
6	Serial Number
7	License Number
8	Date Received
9	Criticality Code
10	Assigned To
11	Quantity

A description of the above eleven items follows.

Equipment Number

The equipment number should be a 6 digit number defined as follows:

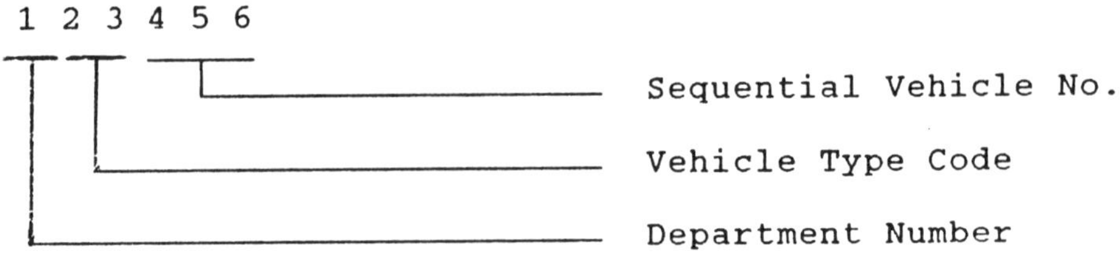

Note: Digits 2 and 3, the vehicle type code, are explained:

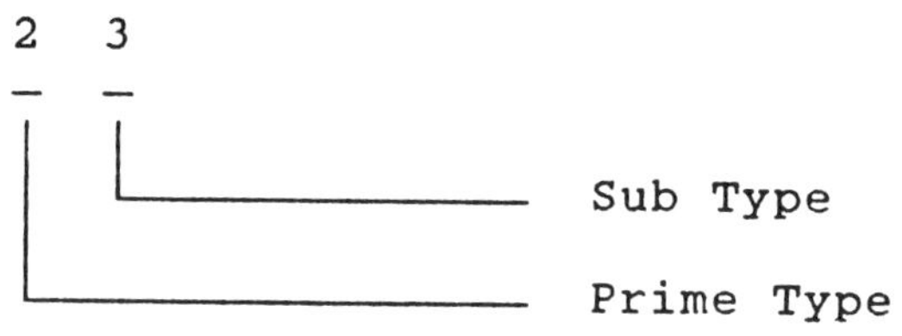

Digit 1, the department number is defined in Figure 5-1.

DEPARTMENT NAME	Expense Acct. No.	Department Number
City Council	101	0
City Manager	102	1
City Secretary	103	2
Purchasing	107	3
Planning & Zoning Commission	111	4
Personnel	112	5
Central Garage & Service Station	116	6
Inspection & Permits	117	7
Police Department	201	8
Fire Department	202	9
Electrical Department	203	A
Traffic Control Devices	204	B
Civil Defense	205	C
Health Department	301	D
Engineering	401	E
Street Maintenance	402	F
Drainage	406	G
Sanitation	409	H
Public Works Administration	410	J
Parks & Recreation	501	K
Pleasure Island Commission	502	L
W&S Purification	1002	M
W&S Distribution	1003	N
W&S Accounting & Collecting	1004	P
W&S Administration & General	1005	Q
W&S Sanitary Sewers	1007	R
W&S Sanitary Sewage Disposal	1009	S

NOTE: Since the letters "i" and "o" might be confused with the numbers "0" and "1", the letters are left out of the alpha characters.

Figure 5-1: Department number.

Digit 2, the prime type vehicle code is defined as follows:

Prime Type Codes

0	Passenger Vehicles
1	Trucks
2	Trailers
3	Heavy Equipment - Self Propelled
4	Heavy Equipment - Trailer Mounted
5	Fire Trucks
6	Boats
7	Motorcycles

Digit 3, the sub-type vehicle code is defined in Figure 5-2. The appropriate prime type code shall be selected first and then the sub-type code from Figure 5-2.

Sub Type Codes for Passenger Vehicles

(Prime Type 0)

1	Compact
2	Intermediate
3	Standard
4	Medium
5	Station Wagon
6	Window Van

Sub Type Codes for Trucks

(Prime Type 1)

0	1/2 Ton Pickup
1	1 Ton
2	2 Ton
3	Truck Tractor
4	Winch
5	Garbage Packer
6	Trash, Self Loading

Figure 5-2: Sub type vehicle codes.

7	Water Tank Truck
8	Dump
9	Dump with Front End Loader
A	Hydraulic Sewer Cleaner
B	Container Handling Compactor
C	Aerial Bucket

Sub Type Codes for Trailers

(Prime Type 2)

0	Garbage Trailer
1	Pound Trailer
2	Utility

Sub Type Codes for Heavy Equipment - Self Propelled

(Prime Type 3)

0	Crawler Tractors
1	Draglines
2	Motor Graders
3	Pulvi-Mixers
4	Hydraulic Backhoes
5	Vactor (Storm Drain Cleaner)
6	Garbage Compactor (Landfill Compactor)
7	Open
8	Open
9	Trenchers
A	Rubber Tired Tractors
B	Gradall - Rubber Tired
C	Gradall - Track Mounted
D	Front End Loader - Rubber Tired
E	Front End Loader - Track Mounted
F	Concrete Mixer
G	Pneumatic Tired Roller
H	Concrete Paving Equipment
J	Street Sweeper
K	Drum Roller
L	Slope Mower
M	Rotor Boom Tractor

Figure 5-2: (continued)

Sub Type Codes for Heavy Equipment - Trailer Mounted

(Prime Type 4)

0	Pneumatic Tired Roller (Compactor)
1	Sheepfoot Roller
2	Lo-boye
3	Routing Machine
4	Air Compressor
5	Asphalt Distributor
6	Drum Roller
7	Concrete Paving Equipment
8	Spreader Box
9	Pump
A	Sprayer
B	Construction Sweeper
C	Sand Trailer

Sub Type Codes for Fire Trucks

(Prime Type 5)

0	Ladder
1	Pumper

Sub Type Code for Boats

(Prime Type 6)

0	All Boats

Sub Type Code for Motorcycles

(Prime Type 7)

0	All Motorcycles

Digits 4, 5 and 6 are the sequential vehicle number. Equipment numbers should be assigned by the garage superintendent, and should be painted or stenciled on each vehicle.

Equipment Description

The equipment description should be the exact word descriptions defining the prime and subtype codes. For example, for vehicle number 803462, the equipment description would be:

"Passenger Vehicle - Standard"

Acquisition Value

Enter the purchase price of the vehicle in dollars.

Make

Enter the equipment manufacturer's name, i.e., "Ford."

Model

Enter the model year, i.e., "1988."

Serial Number

Enter the equipment serial number.

License Number

Enter the license number.

Date Received

Enter the month and year received, i.e., "June, 1987."

Criticality Code

Enter the letter A, B, or C as defined previously.

Assigned To

Enter the name of the person the vehicle is assigned to. If the equipment is pooled, insert the word, "pool."

Quantity

The computer will enter the number of each type of vehicle and the sum total of vehicles per department.

Types of Inventory Lists

There should be three types of inventory lists made:

List 1 - Shall be sorted by vehicle type. This shall be accomplished by computerized sorting of vehicle number digits 2 and 3 (vehicle type codes).

List 2 - Shall be a sorted by department and vehicle type. This shall be accomplished by computerized sorting of vehicle number digits 1, 2 and 3.

List 3 - A listing in numerical sequence of digits 4, 5 and 6 (vehicle sequence number).

Distribution

It should be the responsibility of the garage superintendent to determine the inventory list distribution, both by type of list and person receiving the list.

PREVENTIVE MAINTENANCE SYSTEM - GENERAL

Many plants are using computer software for their fleet management PM program. If a computerized system is desired, the reader is referred to Chapter 23.

The smaller plant may wish to use a manual card type system. This section describes such a system.

The PM paperwork system should consist of three 8 1/2" by 11" card sets prepared for each item of equipment as follows:

- o The Nameplate Data and Scheduling Card (Figure 5-3)
- o The Preventive Maintenance Order (Figure 5-4)
- o Breakdown Maintenance History (Figure 5-5)

It should be the responsibility of the garage superintendent to prepare a card set for each new item of equipment.

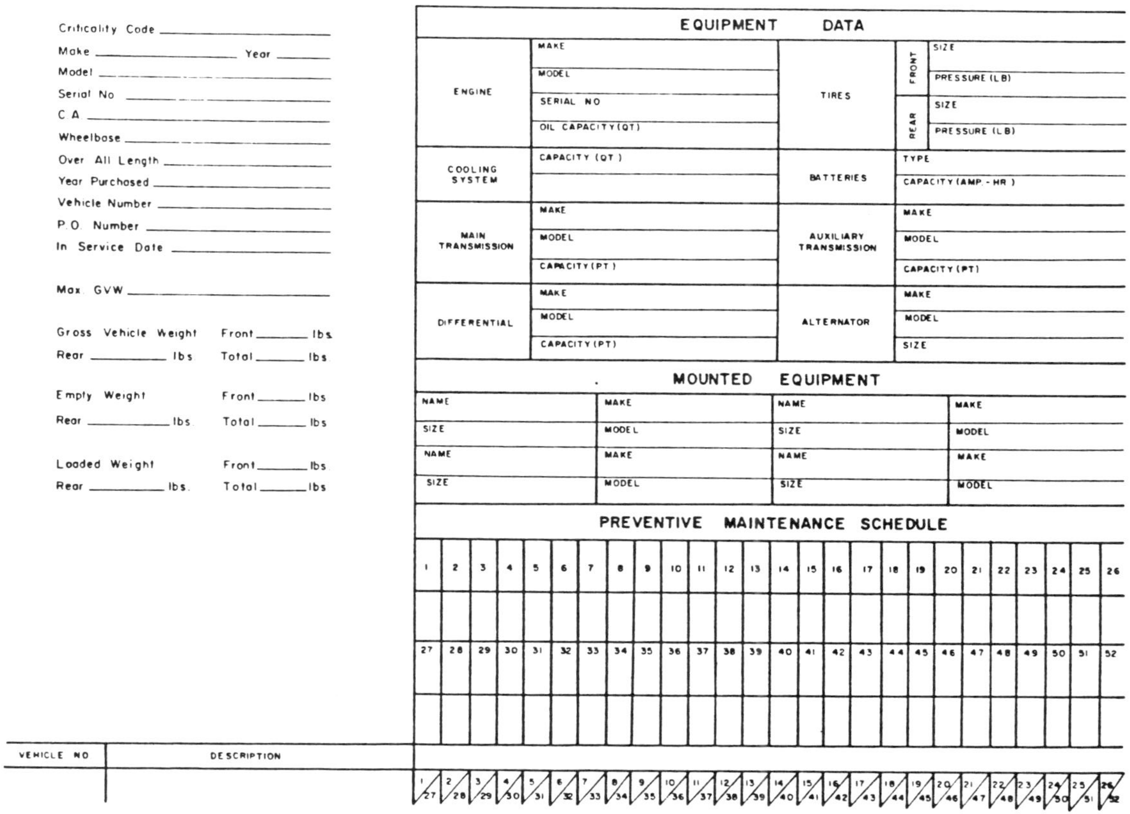

Criticality Code ________
Make ________ Year ________
Model ________
Serial No ________
C.A. ________
Wheelbase ________
Over All Length ________
Year Purchased ________
Vehicle Number ________
P.O. Number ________
In Service Date ________

Max. GVW ________

Gross Vehicle Weight Front ____ lbs
Rear ____ lbs Total ____ lbs

Empty Weight Front ____ lbs
Rear ____ lbs. Total ____ lbs

Loaded Weight Front ____ lbs.
Rear ____ lbs. Total ____ lbs

VEHICLE NO	DESCRIPTION

EQUIPMENT DATA

ENGINE	MAKE	TIRES	FRONT	SIZE
	MODEL			PRESSURE (LB)
	SERIAL NO		REAR	SIZE
	OIL CAPACITY (QT)			PRESSURE (LB)
COOLING SYSTEM	CAPACITY (QT)	BATTERIES		TYPE
				CAPACITY (AMP.-HR.)
MAIN TRANSMISSION	MAKE	AUXILIARY TRANSMISSION		MAKE
	MODEL			MODEL
	CAPACITY (PT)			CAPACITY (PT)
DIFFERENTIAL	MAKE	ALTERNATOR		MAKE
	MODEL			MODEL
	CAPACITY (PT)			SIZE

MOUNTED EQUIPMENT

NAME	MAKE	NAME	MAKE
SIZE	MODEL	SIZE	MODEL
NAME	MAKE	NAME	MAKE
SIZE	MODEL	SIZE	MODEL

PREVENTIVE MAINTENANCE SCHEDULE

1	2	3	4	5	6	7	8	9	10	11	12	13	14	15	16	17	18	19	20	21	22	23	24	25	26
27	28	29	30	31	32	33	34	35	36	37	38	39	40	41	42	43	44	45	46	47	48	49	50	51	52
1/27	2/28	3/29	4/30	5/31	6/32	7/33	8/34	9/35	10/36	11/37	12/38	13/39	14/40	15/41	16/42	17/43	18/44	19/45	20/46	21/47	22/48	23/49	24/50	25/51	26/52

Figure 5-3: Nameplate data and scheduling card.

PREVENTIVE MAINTENANCE ORDER		DATE
EQUIP.	ACCT.	LABOR $
LABOR HRS.	LABOR RATE	ODOMETER HOUR METER

NO.	WORK DESCRIPTION	OK	NG	REMARKS
				EST. HRS. ______

INSPECTED BY:	APPVD. BY:

Figure 5-4: Preventive maintenance order.

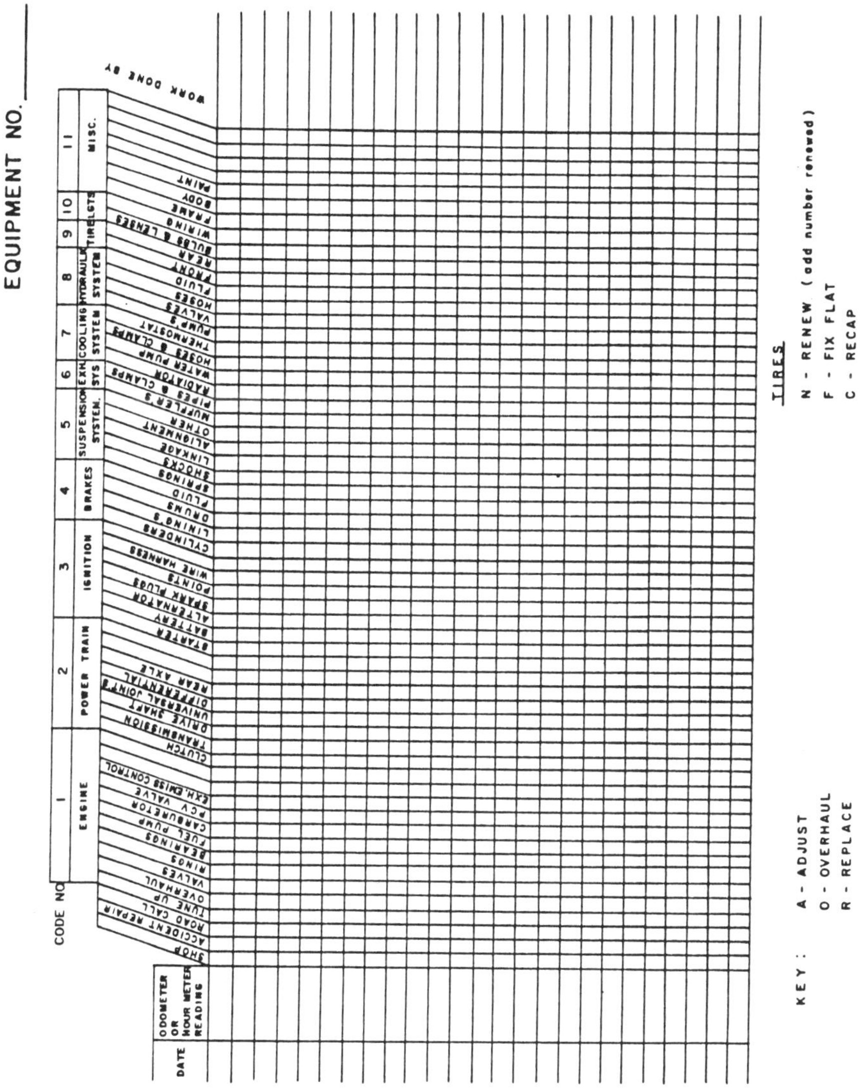

Figure 5-5: Breakdown maintenance history card.

Nameplate data should be entered on the nameplate data card. This card also contains the automatic scheduling device. Scheduling by means of this card is explained in the paragraph on planning and scheduling. The preparation of PM instructions is explained in the following paragraph. The entering of historical breakdown information is described in the paragraph after the next one.

The three PM cards described above, constituting a set of cards for each vehicle, shall be housed in a card holder and cabinet similar to the Acme Visible Record System, or equal.

PM SYSTEM - INSTRUCTIONS

The PM instructions for each vehicle constitute the heart of the PM program. They should be written *only* by the most knowledgeable person in the plant in the field of fleet maintenance. This point cannot be overemphasized. To do otherwise is to invite excessive costs, downtime, production lost time, and criticism of the maintenance program. The writer of PM instructions should give careful consideration to the principle outlined in Chapter 13 on how to write maintenance instructions for optimum results including consideration of:

- o Mandatory versus non-mandatory instructions
- o Positive versus inspection instructions
- o Pyramiding versus non-pyramiding instructions

A separate set of instructions should be written for each *different* type of vehicle. For example, although one set of instructions may suffice for all passenger vehicles, a different set would be written for a bulldozer.

A PM writer, using the form shown in Figure 5-4 proceeds as follows:

a. Fills out equipment number and the account code.

b. Writes the individual instruction in the work description column numbering each work item in sequence.

c. Fills in the estimated labor time required in the space provided at the bottom of the remarks column.

d. Determines the frequency required of work items.

e. Selects the frequency code from the following code table:

Frequency	Code
Annual	A
Semi-Annual	S
Quarterly	Q
Monthly	M
Semi-Monthly	SM
Weekly	W

f. Inserts the frequency code in the box provided next to the date box in the upper right hand corner.

g. Special spare parts required for the PM tasks are listed on the bottom of the form.

When the work is actually performed, the mechanic does the following:

a. Checks the "OK" box if the inspection reveals no problem or if a positive instruction, i.e., "replace belt," is carried out.

b. Checks the "NG" column if the item to be checked is bad, i.e., "inspect belts." If he replaced or adjusted the belt in this case, he would spell out the specific action in the remarks column.

c. Enters actual labor hours in the "Labor Hours" box on top of the form.

The foreman checks the work and signs the "Approved By" box. The office clerk then checks the entries and completes the labor rate and labor dollars column.

The office clerk makes a copy of the completed PM form and files the copy in the equipment folder and sends the original to accounting for its records.

PM SYSTEM - HISTORICAL RECORDS

Breakdown repairs are recorded on the garage work order (Figure 5-6). In order to measure the repetitiveness of the type of breakdown, information from the garage work order (Figure 5-6) shall be recorded on the breakdown maintenance history card (Figure 5-5).

For example, if a vehicle broke down on the road because of a battery failure and if the battery was replaced, the following entries would be made on the breakdown card by the office clerk:

- o Date of breakdown.
- o Odometer or hour meter reading.
- o Check road call box.
- o Place an "R" for replace (see key on card) in the box provided under ignition - battery.

The clerk, when making the entry, will look at the entire card observing the repetitiveness of failures. If three or more failures occurred on the same item, the clerk shall notify the garage superintendent who shall analyze the problem to see what can be done to eliminate the failure in accordance with the method described under the maintenance analysis paragraph.

A separate folder with the number and name description shall be kept for each item of vehicular equipment. In it shall be filed in chronological sequence:

- o Completed PM orders
- o Completed garage work orders
- o Outside repair orders
- o Pertinent correspondence
- o Purchase order and specs of vehicle

EQUIP.	**GARAGE WORK ORDER**	DATE
ACCT.		LABOR $
REPAIR HRS.	DOWNTIME HRS.	ODOMETER HOUR METER

WORK REQUESTED	
REQUESTOR:	

WORK DONE DESCRIPTION	WORK DONE BY:	RATE	REPAIR HOURS	
			EST.	ACT.

MATERIAL USED QTY.	DESCRIPTION	VENDOR	INVOICE NO.	P.O. NO.	PRICE $

DOWNTIME	OUT OF SERVICE		BACK IN SERVICE	
	DATE:	TIME:	DATE:	TIME:

Figure 5-6: Garage work order.

In summary, the historical record system shall consist of three items:

- o Management information system
- o Breakdown maintenance history card
- o Equipment folder

PM SYSTEM - MAINTENANCE ANALYSIS

General

Maintenance analysis is an analysis of the existing preventive maintenance program. It tries to determine if we are operating at the optimum points shown in the graphs of Figures 1-2 and 2-14. The maintenance analysis program utilizes three systems as follows:

<u>Broad Band Control</u>

This is the computerized management information system described in Chapter 29.

<u>Intermediate Band Control</u>

This is the breakdown maintenance history card.

<u>Detailed Control</u>

This is contained in the equipment file.

Fundamentally, we are looking for the reasons for equipment failures so that we can take action to prevent future failures and thus reduce downtime. But the maintenance staff cannot analyze everything! There must be some convenient method of surfacing only problems of significance where corrective action is both possible and desirable. This system is designed with this point mind.

Maintenance analyses should sort out, analyze and eliminate important problems as suggested in the following steps:

a. Categorize by criticality.

b. Examine deviations from standards in broad band feedback report.

c. Isolate repetitive failures on breakdown history card.

d. Locate detailed cause of failure from equipment file.

e. Analyze failure and take corrective action.

An examination of these steps follows.

Categorize by Criticality

Equipment must be coded by criticality in order to select which equipment items are to be given first attention in the analysis effort.

Examine Deviations from Standards

In Chapter 29, a broad band management information system is described that provides a monthly single line report for each vehicle segregated by department and vehicle type. "Broad Band" refers to the fact that the information is "broad" in scope as opposed to "detailed."

Examples of information furnished in this report that can be used for maintenance analysis are:

- Monthly and since acquisition costs for preventive maintenance, breakdown maintenance, and operation.
- Monthly breakdown maintenance hours expended.
- Monthly downtime hours.
- Percentage breakdown maintenance to total maintenance costs. (A measure of the success or failure of the PM program.)
- Ratio of downtime to maintenance hours. (A measure of the length of repair time relative to downtime.)
- Miles per gallon. (A measure of engine efficiency.)

It shall be the garage superintendent's responsibility to establish the standards for each of these parameters and to observe the monthly values of each parameter relative to the standard to determine deviations beyond allowable tolerances. For example, if a standard of 15 miles per gallon were established on passenger vehicles and a particular vehicle showed a figure of three miles per gallon, then the report would surface a problem with engine efficiency that should be investigated further by means of the intermediate and detailed feedback systems.

In other words, the broad band information system is a means for isolating problem areas, but not for surfacing the problem in detail.

Isolate Repetitive Failures

The breakdown history card is used to surface repetitive failures. It will be referred to under two circumstances:

a. When the broad band system discloses a deviation from standard for a particular vehicle.

b. When the clerk, in making new entry on the history card, observes three or more failures of the same kind on a specific vehicle.

Determine Specific Failure Cause

When a repetitive problem is surfaced on a specific vehicle, the analyzer must thumb through the PM orders and the Garage Work Orders to determine the cause of the failure.

Analyze Failure and Determine Corrective Action

The first step in analyzing the failure is to determine the cause category as shown in Figure 2-1. Broad solutions to the problem are also shown in this figure. Most of the corrective actions shown in this exhibit are self evident. A detailed discussion of the interaction between PM and Breakdown Maintenance and how to go about eliminating correctable repetitive breakdowns is discussed in Chapters 9 through 13.

BREAKDOWN REPAIRS

General

All repair work not called for on the Preventive Maintenance Order (Figure 5-4) shall be recorded on the garage work order (Figure 5-6). An explanation of the latter form follows:

When the driver of a vehicle experiences trouble with a vehicle, he shall make out a garage work order and shall fill out the following items:

- o Equipment Number
- o Date
- o Work Requested
- o Requestor
- o Date and time of service
- o Odometer reading

He should be as specific as possible in the work requested column. For example, if the vehicle won't back up, he should write:

Work Requested

Vehicle won't back up - please fix!

An unsatisfactory statement might be:

Work Requested

Vehicle doesn't work right

Unit N.G.

Please Fix

The garage foreman, when receiving the vehicle, checks over the items filled out by the driver for correctness. He modifies the work requested box if it doesn't accurately describe the problem. He then determines the work to be done and describes this in the work done box. In this case the work description might be:

Work Done

Check automatic transmission for failure to back up.

He then puts his estimate of the manhours required to perform this work and inserts the number in the repair hours estimate column.

The foreman then assigns the work to a mechanic. The mechanic then inspects the transmission and decides that it must be dismantled for inspection to determine the cause.

Upon dismantling the transmission, he discovers a broken back up spring which he replaces. He decides in addition, since the unit is dismantled, that he will renew all gaskets and seals, clean the unit and renew the transmission oil.

Upon completion of the job, the mechanic will make the following entries:

Work Done

Replace broken back up spring

Renewed all gaskets and seals

Cleaned, reassembled and installed

Renewed transmission oil

The mechanic will enter the actual time spent on the job and date and time the unit was ready for service. The completed work done portion of the work order will look as follows after the mechanic has completed his work:

Description	Work Done By	Rate	Est.	Act
Check automatic transmission for failure to back up.	J. Jones	20	4	4
Replaced broken back up spring. Renewed all gaskets and seals.	J. Jones		8	10
Cleaned, reassembled and installed. Renewed transmission oil.	A. Smith	15	4	4
Total			16	18

The foreman will then place his initials opposite the actual hours signifying that he has checked the reasonableness of the actual hours compared to his estimate. If there is a *large* deviation, he shall determine the cause and take corrective action if necessary.

The clerk shall then:

Fill out the remaining boxes which are the labor rates, total repair hours, downtime hours, the labor dollar, and the account code.

Transfer information to the historical record card. Duplicate the order and send one copy to accounting. File one copy in the equipment file in chronological order.

There are three fundamental purposes for the garage work order. These are:

a. To provide statistical input into the accounting system and the transportation management information system.

b. To surface the cause of failure.

c. To provide labor and performance measurement information (estimate versus actual).

d. To facilitate planning and scheduling of garage labor.

The processing of information to accounting, and performance measurement has been described in this paragraph.

The use of the garage work order for planning and scheduling purposes is described in the paragraph on planning and scheduling. The following paragraph discusses further the "art" of surfacing the cause of failures.

Surfacing the Cause of Failure

Experience has shown that accurate reporting of the cause of failure is always a *very big problem*!

Mechanics must be taught very clearly what the object is and that is to record the exact cause of the failure, if known.

Using the previous example of the vehicle which wouldn't back up, typical unsatisfactory statements are:

o No statement at all.

o Unit fixed (this says nothing).

- o Overhauled transmission (this does not indicate what was wrong).
- o Replace back up spring, oil, gaskets, reverse gear, miscellaneous nuts and bolts (which of these caused the failure?).

It shall be the responsibility of the garage foreman and the superintendent to continuously audit and check the work done box until the feedback information is meaningful and the mechanics realize that management is serious in its desire to obtain accurate feedback information.

SERVICE WORK

Operator Scheduled Maintenance

Operators and drivers of fleet equipment must be required to inspect the unit to which they have been assigned at the start and completion of their operating shifts. Defects discovered during such inspection, or during operation of the equipment, should be noted on the existing garage work order form and reported at the completion of the regular shift. Obviously the operator should stop immediately if he observes a deficiency that could cause the equipment to become damaged or unsafe. Department heads have the responsibility of monitoring this inspection program.

Service Station Type Servicing

Services that must be performed during each fueling of an equipment unit include:

- o Check oil level in crankcase.
- o Check tires for proper inflation and condition.
- o Check electrolyte level in battery.
- o Check coolant level and antifreeze.

This work is ordinarily done at the plant service stations, however, when equipment is brought to outside service stations for fueling, the equipment operator is responsible for performing these services or assuring that this work is performed by others.

People (drivers) who have this responsibility can be divided into two categories: those who care about maintenance and those who do not. Because of the people in the latter category, we must constantly remind drivers of their responsibility in this regard.

The point to be made is that it *is not* their car to do with at they please, it is plant property to be cared for as directed and that the above simple check which is designed to reduce vehicle downtime and maintenance cost is a must. This directive may be requiring them to perform work they are not used to doing on their own automobile. It is for this reason that periodic publicity is desirable.

MATERIALS

Although most plants have a well stocked storeroom for manufacturing operations, it is frequently found that there are no spare parts carried in the garage and that most are obtained from local vendors. This is not a good situation particularly if the supply house is far away and can lead to unnecessary vehicle downtime and non-productive labor hours. Full use should be made of the material management techniques discussed in Chapter 19.

Many plant service stations have only the following material items stocked:

- Gasoline
- Oils
- Greases
- Tires
- Tubes
- Anti-Freeze
- Transmission Fluid

The following suggestions are made for the average plant service station:

- Stock fast moving items for categories A and B vehicles in the garage.
- Utilize inventory management techniques described in Chapter 19.
- Monitor usage and eliminate slow moving items from stock.

- o Monitor purchases of non-stock items and add fast moving items to stock.

- o Make use of periodic free inventory analysis offered by most supply houses for automotive parts.

An extra special effort should be made in determining stock requirements for class A vehicles, particularly those required to support the PM program. These parts should be "kitted" well in advance of the scheduled PM task.

Kitting, as previously discussed in Chapter 19, is the act of obtaining the materials needed for a particular task in advance of the task and putting them in a bin or box set aside for the task. What we are looking for in order to increase PM job productivity is to have the following elements of the job available before the job starts:

- o Manpower
- o Instructions
- o Materials
- o Tools and Equipment

PLANNING, SCHEDULING AND CONTROL OF WORK

General

The methods used for planning, scheduling and control of garage work vary with the type of work. These work types are:

- o Fixed Assignments
- o Preventive Maintenance
- o Non-Productive Time
- o Breakdown Maintenance

The following paragraphs describe how each of these work types should be handled:

Fixed Assignments

Fixed assignments are the handling of routine repetitive tasks such as those performed by the:

- o Janitor
- o Office Clerk
- o Servicemen

Control of these functions is obtained by:

- o Listing and defining the parameters of the routine repetitive tasks performed.
- o Determining the daily quantity of these tasks.
- o Time studying each task.
- o Establishing a standard time for each task.
- o Multiplying the standard by the daily number of occurrences.
- o Assigning each man a day's worth of work as calculated above.
- o Spot sampling performance to standard on a periodic basis.

The above functions can be illustrated by a hypothetical janitor's job. Let us assume that he performs the following tasks:

Task Description	Quantity	Std Hrs/Task	Tot Hrs
Clean Toilet Bowls	10	.15	1.5
Clean Lavatories	10	.15	1.5
Clean Toilet Floor & Walls	4	.50	2.0
Replace Paper	10	.05	.5
Clean Garage	2	.15	1.5

In the case of most janitors' tasks, the standard times can be obtained from textbooks. Some tasks, such as cleaning the garage, must be time studied and a standard established.

In the case of the garage clerk, all of the tasks should be time studied. An integral part of each time study is the definition of how the task is to be performed. For example, the time required to clean the garage could conceivably vary from 5 minutes to 2 hours, depending upon how the job was defined.

Although the above seems like a lot of work, it really

isn't since it need only be performed once. After this one shot exercise, the balance of the control function consists of periodically auditing compliance of the worker to the prescribed task.

The above type of control is invaluable in determining when to increase or decrease the labor force. In the case of the janitor's job described above, he has been loaded up to 7 hours only. If a new toilet room were added consisting of one toilet bowl, lavatory, paper dispenser, walls and floors, it would add a work load of .85 hours, which would still be inside of 8 hours. Therefore no extra man would have to be added.

Preventive Maintenance

The first task in scheduling PM work is to balance the workload as evenly as possible throughout each of the 52 weeks in the year. This is achieved by use of a PM load balancing chart. The format of this chart shall be as follows:

LOAD BALANCING CHART

Equip No	Weeks																
	1	2	3	4	5	6	7	8	9	10	11	12	13	....	49	50	51	52
001	7				1				1				3		3			
Total	7				1				1				3		3			

Annual = 4 hours
Quarterly = 2 hours
Monthly = 1 hour

Every vehicle in the PM program would be listed on this load balancing chart.

For illustration purposes, let us assume that at the end of a particular week, the superintendent has completed instructions for 3 passenger vehicles and 2 gradalls. The labor estimates for each of these vehicles are:

Passenger Vehicles	PM Manhours		
	M	Q	A
001	1	2	4
002	1	2	4
003	1	2	4

M, Q, and A refer to the frequency codes previously described.

Gradalls	PM Manhours		
	M	Q	A
006	1	6	20
007	1	6	20

The first thing to recognize is that each frequency possesses various degrees of load spreading capability. For example, a monthly inspection can be performed on the 1st week of the month, the 2nd week, the 3rd week, or the 4th for 4 choices. Similarly the other frequencies have the following number of choices:

W	1
M	4
Q	13
S	26
A	52

With this point in mind, the 5 new vehicle loads are added to the previously described chart being careful to spread the load as evenly as possible.

Monthly inspections will always include Weekly's, Quarterly inspections will always include Monthly's, Semi-Annual inspections will always include Quarterly's and Annual inspections will always include Semi-Annual's.

As each new PM inspection is written, it shall be added to the charge and the totals per week updated. In order to help visualize the load balancing operation, a graph shall be maintained of the total column as shown in Figure 5-7 . This graph and the workload balancing chart shall be maintained by the garage clerk. The next task is to set up the individual vehicle scheduling cards (see Figure 5-3).

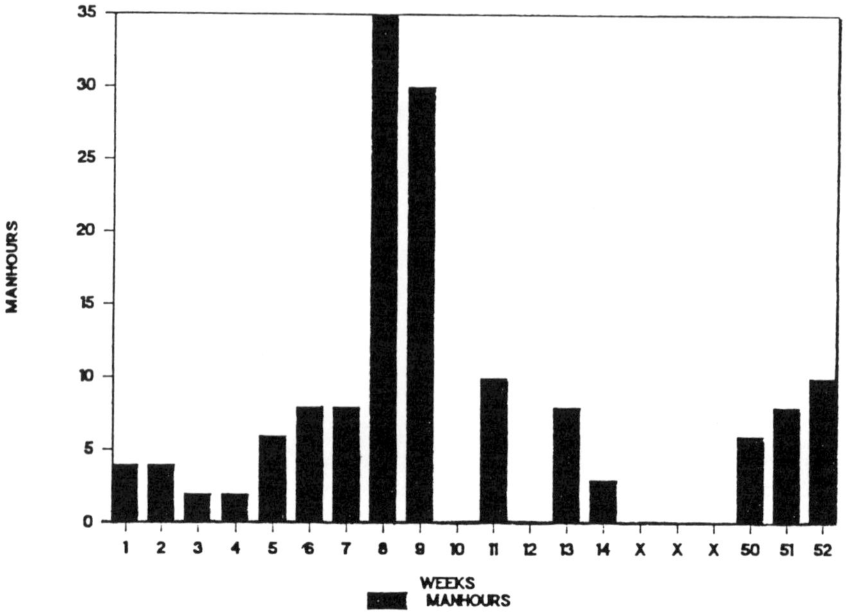

Figure 5-7: PM workload graph.

The information contained on the load balancing chart is entered into the scheduling card (Figure 5-3) by putting the frequency codes where desired. The scheduling operation is started by placing the sliding colored scheduling tab that runs along the bottom of the card to the first scheduled event which in this case is on the first week of the year. Once a week on Friday afternoon, the garage clerk shall scan the PM card file to see which colored tabs are showing for that particular week. The vehicles scheduled for that week shall have their PM orders pulled and xeroxed. The master PM cards shall then be returned to the equipment PM file.

The sequencing of events shall occur as follows for hypothetical week number 5:

- On Friday of the 3rd week, pull cards on vehicles due the 5th week.
- On Monday of the 4th week, send notice to the department heads announcing vehicles due for PM on the 5th week.
- On Friday of the 4th week, issue PM work orders for the 5th week to foremen.

- o Perform scheduled PM work sometime during the 5th week.

Non-Productive Time

Non-productive time includes such items as:

- o Vacations
- o Sick Leave
- o Coffee Time
- o Idle Time (time available for work, but not working)
- o Miscellaneous Absences
- o Union Business

It shall be the responsibility of the garage superintendent to obtain whatever statistical information that is required for him to have an approximate idea of the average number of manhours per day charged to the above non-productive hour items. This knowledge is necessary for him to estimate the manhours available for productive work. Without this approximation, he will be unable to intelligently plan and schedule his daily work.

Breakdown Maintenance

There are two important reasons for planning and scheduling:

- o To provide each worker with a day's worth of work each day and to see that he performs a day's worth of work each day.
- o To provide as far as is possible prior to the start of each job the following:

  Manpower

  Instructions

  Materials

  Tools and Equipment

The above two items are essential to high labor productivity. The items mentioned previously, i.e., fixed assignments, PM and non-productive time, form the base

SHOP SCHEDULE ____________ GARAGE | Date ____________

Capacity:

1. ______Men x 8 Hours Each = ________ Hrs.
2. Anticipated Overtime ________ Hrs.
3. Total Capacity(1 + 2) ________ Hrs.

Fixed Assignments:

4. ____________________ ________ Hrs.
5. ____________________ ________ Hrs.
6. ____________________ ________ Hrs.
7. ____________________ ________ Hrs.

Unavailable Time:

8. Vacations ________ Hrs.
9. Sick Leave, Absences ________ Hrs.
10. Other (Specify) ________ Hrs.

11. Hours Available for Schedule (3 minus 4 thru 10) ________ Hrs.

Schedule:

Item No.	Vehicle No.	Work Description	Mechanic(s)	Hours		
				Est.	Act.	Variance
1.						
2.						
3.						
4.						
5.						
6.						
7.						
8.						
9.						
		Totals				

Figure 5-8: Shop schedule form.

workload of the garage. The time left over is spent on breakdown repairs.

Breakdown maintenance work is unanticipated and highly variable and is managed on a backlog control basis. The meaning of backlog control will be explained in the following paragraphs.

The shop schedule form shall be used by all garages and service stations (Figure 5-8). The foreman in charge of each garage or service station shall make out one of these forms each month filling out items 1-11. This form shall then be duplicated and used on a daily basis to plan the day's workload of Preventive Maintenance and Breakdown Repairs. The hours shown in column 11 are the available hours for PM and breakdowns.

The schedule portion of the form shall be used to determine the issuance of a day's worth of work for that day to measure the mechanic's performance.

Each day there will be a known input of work since both PM and breakdown work will be estimated. The theoretical output of work is the available hours shown in item 11. If the input exceeds the output, a backlog of jobs will be created. If the opposite occurs, the backlog will be reduced.

If the backlog reduces to zero or close to it, excess labor capacity is indicated and a reduction in staff is suggested. Conversely, if the backlog continuously rises, an increase in staff or overtime or use of outside contractors is indicated.

The measuring and control of these items is called backlog control and is the responsibility of each garage head and the garage superintendent. This subject is discussed in more detail in Chapter 16.

27

Utilization

GENERAL

Proper utilization is encouraged by:

- o Having the correct number of vehicles (too many vehicles causes low utilization).
- o Having intermittently used equipment under dispatching control.
- o Selecting the correct vehicle for the job.
- o Recording mileage or operating hour information and publishing this information on a periodic basis for utilization analysis purposes.

Comments on these items follow.

EQUIPMENT SELECTION

High utilization of a vehicle is encouraged by selecting the correct type and quantity of vehicles for the particular job that it must perform. For this reason careful consideration should be given to equipment selection.

Suggestions for determining correct equipment selection are given in Chapter 28.

CONTROL OF INTERMITTENTLY USED VEHICLES

General

Many heavy equipment items are logically assigned to specific individuals or operating groups on a permanent basis. An example of such an item is a street sweeper. This item is operated by one man who uses it continuously all day, every day. On the other hand, certain heavy equipment items required by several departments are used infrequently or intermittently. An example of this is a bulldozer or backhoe. Full utilization of such equipment can only be assured by pooling.

Accordingly, intermittently used heavy equipment items shall be identified by the garage superintendent with the aid of the various department heads. These items shall be put in a pool managed by the garage superintendent. Operation of this pool is described in the following paragraph.

Heavy Equipment Pool Operation

A vehicle dispatch board (magnetic type), shall be prepared for controlling all heavy equipment pooled vehicles. The board will be similar to that shown in Figure 5-9. An explanation of this board follows. The top horizontal heading of the board lists all the types of pooled vehicles, i.e., trenchers, backhoes, bulldozers, etc. The vertical columns under each heading are divided into two parts, "in use" and "pool." A magnetic type plaque card shall be made for each vehicle imprinted with the vehicle number as follows:

653022

All vehicle number plaques shall be green with black letters.

In addition an adequate number of blank yellow plaques shall be made available to show the job number and location of the vehicle assignment as follows:

	COMPRESSOR TRUCKS	TRENCHERS	BACKHOES	HYDRO PAVING BREAKERS	DUMP TRUCKS	TRAILERS	SEDANS	STATION WAGONS	ETC.
IN USE									
POOL									

Figure 5-9: Vehicle dispatch board.

Job No.	Location
1234	SE corner of 6th and Grand Streets

When a vehicle is assigned for use, the vehicle plaque plus the job identification plaque shall be placed under the appropriate classification column in the "in use" portion of the column.

Equipment not being used shall be identified by placing the equipment number plaque in the "pool" section of the board. This shows that the equipment is located in the home base parking area.

When a vehicle in use in the field has completed its function and use of it is no longer contemplated for the remainder of the day, the driver shall radio the dispatcher and inform him of this change in status. The dispatcher shall thereupon remove the vehicle *and* job identification plaque and place it in the "pool" portion of the column signifying that the equipment is available at the location shown. If the equipment is subsequently returned to home base, the driver shall notify the dispatcher who shall remove the location plaque opposite the vehicle number signifying that it is available at home base.

Ordinarily, equipment requirements shall be filled from the home based "pool" vehicles. However, if a particular type of home based pooled vehicle is unavailable, a "pooled" vehicle on a job site may be utilized. The decision on the economics of such a move shall be made by the superintendent or the person he designates to perform this function.

Each day at approximately 3 PM, the dispatcher shall record on the form shown in Figure 5-10, the number of "pool" vehicles available as indicated on the dispatch board. This information can then provide a statistical basis for making decisions on:

a. Acquisition

b. Disposals

c. Utilization Improvement Methods

The statistics may very well show that the use of some vehicles is seasonal, which may lead to justification for equipment rentals. This analysis will be facilitated by the vehicle cost report outlined in Chapter 29, "The Transportation Management Information System."

MEASUREMENT OF EQUIPMENT USE

All equipment shall be provided with either an odometer for measuring mileage or an hour meter as appropriate. Once per month the driver of the vehicle shall fill out a form showing the miles or hours operated per month. The report shall be made out in duplicate, one copy to be retained by the driver and one transmitted to accounting for submission to the computer operator. The following form shall be used for this purpose:

```
MONTHLY VEHICLE USE REPORT

Equipment NO.____________________
Date ________ Driver_____________

Odometer/Hour Meter
Reading This Month _____________

Odometer/Hour Meter
Reading Last Month _____________

Miles/Hours Operated __________
```

The monthly report published as part of the Transportation Management Information System described in Chapter 29, will contain the following information useful for utilization analysis purposes:

- o Miles or hours operated per month per vehicle.

UNUTILIZED VEHICLE REPORT MONTH OF ____________

DAY OF MONTH		TRENCHERS		DUMP TRUCKS		BACKHOES		TRAILERS		ETC.	
		H	J	H	J	H	J	H	J	H	J
1											
2											
3											
4											
5											
6											
7											
8											
• •											
• •											
• 31											
MONTHLY AVERAGE PER DAY											

H = HOME BASED VEHICLES J = VEHICLES LOCATED ON JOB SITE

Figure 5-10: Unutilized vehicle report.

- o Miles or hours operated since acquisition per vehicle.
- o Average mileage or hours per month operated per vehicle since acquisition.

It is suggested that each department head, with the help of the garage superintendent, establish standards for monthly usage. Vehicles that show up in the monthly report as being too far out of line from this standard should be subjected to a special analysis as described in the following paragraph.

ANALYSIS OF UNDERUTILIZED VEHICLES

Vehicles that indicate too low a utilization shall receive a special spot sampling analysis. This analysis should be conducted as a one time exercise over a period of 1 to 4 weeks or whatever time is necessary to arrive at a conclusion of what is necessary to correct the poor utilization.

The utilization percentage is defined as the "mobile" percentage in the following tabulation:

Use Category	Percent
Mobile	?
Immobile	?
In shop awaiting parts	?
In shop being worked on	?
Idle	?
Total	100 %

A definition of the terms used in this tabulation follows:

Mobile - Vehicle is being used and is moving

Immobile - Vehicle is being used and is not moving

In Shop - Vehicle is in shop for repairs

Idle - Vehicle is not assigned, used or moving

Collection of this information requires that the driver fill out a form as shown in Figure 5-11. This information is then summarized on a form such as Figure 5-12. It is recommended that utilization forms be filled out on a daily basis by the driver of the vehicles under study.

DAILY UTILIZATION RECORD
MOTOR VEHICLES

Date________ Normal Work Day:_____ Hours Normal Work Week:______ Days Vehicle No._______

24 Hour Clock Time		Mileage		Record of: - Each Destination. From and To - Lunch Perid. In Shop, Idle	Minutes				Initials
Begin	End	Odometer	Miles		Mobile	Immobile	In Shop	Idle	
Total Miles				Total Minutes					
Total Hours									
% of Total Work Day									

Figure 5-11: Daily utilization record for motor vehicles (front).

INSTRUCTIONS

1. ACCURACY. It is imperative that entries be accurate.

2. DAILY RECORD. At least one form will be prepared for each vehicle each day; if a vehicle is idle or in shop all day, one entry will be made; if trips during a single day exceed the number of lines on the form, use a second form.

3. TIME. The number of minutes between the beginning and end 24 hour clock times on each line must be accounted for among the four columns; mobile, immobile, in shop, idle. Lunch periods must be identified by separate entry so that the hours in "Normal Work Day" and "% of Total Work Day" will not be distorted by lunch periods.

4. FROM AND TO. These entries should be easily identified by personnel in the locality where the vehicle is used, such as the name of the city, town, village, "Operations Center", "Building No.", or "General Office Building".

5. CALCULATIONS. The "% of Total Work Day" (bottom of form) should be computed by (a) the assignee-operator of the regularly assigned vehicle, or (b) clerical or administrative personnel convenient to a group of vehicles.

Figure 5-11 (cont'd): Daily utilization record for motor vehicles (rear).

SUMMARY OF VEHICLE UTILIZATION
(By Vehicle Types)

Month ____________________

Vehicle Number	Odometer	Miles Operated	Total Hours /1 Available	Mobile		Immobile		In Shop		Idle	
				Hours /2	% /3	Hours	%	Hours	%	Hours	%
				SEDANS							
				STATION WAGONS							
				ETC.							

/1 Total of hours per month vehicle should be available — number of work days times daily hours — (less lunch periods).

/2 Hours.

/3 Total available hours (/1) divided by hours (/2).

Figure 5-12: Summary of vehicle utilization by vehicle types.

Typical recommendations for correcting low mileage might be as follows:

- o Driver to use existing pool car. His car to be deleted from inventory.
- o Driver's car to be pooled to increase utilization.
- o Downtime on vehicle is excessive, causing low mileage. PM program to be implemented.
- o Car to be exchanged with unit having very high mileage.

A periodic utilization report may show a favorable miles per vehicle average for a *group* of vehicles, however, the average may be disguising low mileage on specific vehicles. Replacement of vehicles is determined both on the basis of age and mileage. It is possible with a replacement policy of this type to sell some vehicles at very high mileage and low age and others at very old age and low mileage. Both of these extremes will very likely cause unnecessary maintenance costs. The answer to this dilemma is rotation between the high and low mileage units of the same type to balance age and mileage accumulations.

28

Justification, Acquisition and Disposal

GENERAL

There are many facets to the problem of controlling the acquisition of vehicles. Some of the main things to consider are:

- Design of adequate justification procedures to prevent purchase of unnecessary vehicles with attendant increased operating expense.
- Design of an adequate planning system that insures purchase and delivery of vehicles when needed.
- Purchasing procedures designed to obtain specified equipment at the lowest or "optimum" cost, all factors considered.
- Design of standard specifications for the purpose of insuring purchase of the best vehicle for the job *and* for reducing the number of types of vehicles to the minimum.
- A purchasing policy that not only provides for the purchase at optimum cost, but also gives consideration to keeping the numbers of make/type combinations to a minimum.

- o Obtaining reasonable value for retired vehicles.

A discussion of these factors follows:

JUSTIFICATION

General

The following procedures shall be used for justifying new and replacement equipment. These systems are divided into the following categories:

- o Standard Justifications
- o Methods Improvement (vehicle purchase to improve an old method of doing job)
- o New Requirement
- o Replacement

A discussion of these categories follows:

Standard Justification

Throughout the plant there exists positions or functions that automatically justify the purchase of a vehicle. An example of this case is a floor sweeper driving a floor sweeping machine. (It is your intention when hiring a new floor sweeper that he will get a machine and spend all of this time sweeping floors with the machine.)

It is recommended that a list be compiled of all these positions and functions and that someone be assigned to writing standard justifications for each of these positions. After the list has been carefully screened and approved, it would be established policy and distributed to all department heads.

Thereafter, when justifying a vehicle falling within one of the standard justifications, it would only be necessary to reference the standard justification number. It would be up to the person approving the acquisition to make certain that the standard justification does, in fact, apply.

Methods Improvement (Vehicle Purchase to Improve an Old Method of Doing a Job)

Whenever a methods improvement is contemplated which involves purchase of a new vehicle, the justification should include the following:

- o Annual cost of the old method
- o Annual cost of the new method
- o Annual savings
- o Investment cost
- o Payoff time

This is standard procedure for justifying a methods improvement that applies equally well to improvements involving vehicle acquisition.

New Requirements

New requirements fall into two categories:

- o An expansion of an existing requirement. . . For example, let's say that one more floor sweeper is hired. This could probably be handled by a standard justification. However, if there was no standard justification, then the justification should include an analysis of the utilization of the existing vehicles used for the same purpose and an explanation of why an additional vehicle is required.
- o A bona fide new requirement . . . In other words, the purchase of a vehicle for reasons never before used. This type of justification should include:
  - Justification for purchase
  - Impact of not purchasing
  - Alternate recommendations

    For example:

    Leasing

    Renting

Personal car use

Taxi

Other

Replacement

Justification for replacement of a vehicle represents the majority of cases of new vehicle acquisition. There are three factors to be considered:

- Rejustification.
- Utilization analysis of existing vehicles used for the same purpose.
- Economic analysis of old versus new vehicle annual costs and payoff time.

Concerning rejustification of the need, the facts should present the original justification and statements to the effect that the function or position has not changed. Change of job requirements throughout the years is not uncommon. These changes *may void* the original justification.

Concerning utilization analysis, this analysis should study the utilization percentage of the existing fleet used for the same purpose and demonstrate statistically that a replacement is justified.

Concerning the economic analysis of the old versus the new vehicle, the following method of analysis is recommended:

For the old vehicle the following cost information should be collected:

1.	Blue book value last year	$ ________
2.	Blue book value this year	$ ________
3.	Depreciation cost (line 1 - line 2)	$ ________
4.	Maintenance cost	$ ________
5.	Operating cost	$ ________
6.	Total annual cost (lines 3 + 4 + 5)	$ ________

The same information must be gathered for the new vehicle as follows:

1.	Acquisition cost (from estimate)	$ ________
2.	Resale cost at end of estimated standard life from blue book	$ ________
3.	Net investment (lines 1 and 2)	$ ________
4.	Average annual cost of similar units from existing record including maintenance and operation	$ ________
5.	Estimated annual average increase in maintenance and operation costs over the anticipated life span of the new vehicle (estimate an average annual increase of 2.5% modified by known changes, such as miles/gallon	$ ________
6.	Estimated annual total cost of depreciation (line 3 divided by estimated life)	$ ________
7.	Estimated maintenance and operations cost (lines 4 plus 5)	$ ________
8.	Total annual cost (lines 6 plus 7)	$ ________

The difference between line 8 of the new unit and line 6 of the old unit represents the estimated savings or increased cost. The estimated savings divided by the net investment (line 3) times 100 equals the estimated payoff rate. Payoff rates of between 10 - 15% should receive favorable consideration.

The analysis described in the previous paragraph can be described graphically as shown in Figure 5-13. Comments on this exhibit follow:

The three annual costs shown in the exhibit are capital shrinkage (annual difference in blue book values), operation costs and maintenance costs. These curves have typical trends as shown and combine to give a total cost curve as shown. Notice that the total cost curve initially drops due to the overriding influence of the capital shrinkage curve which has an initial high decreasing slope. This trend finally

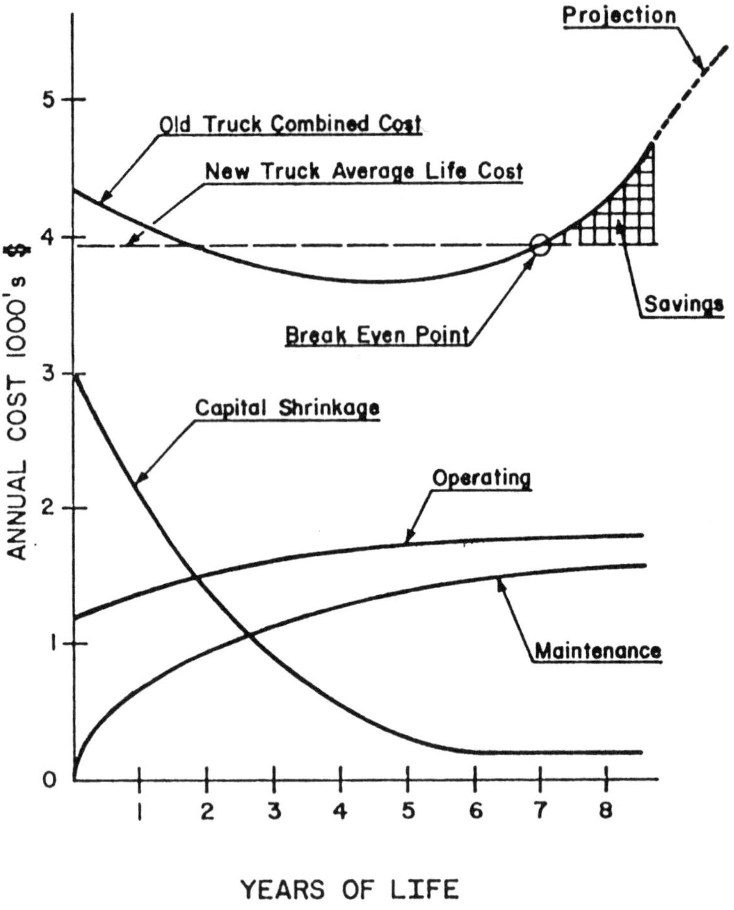

Figure 5-13: Replacement analysis

reverses itself and starts to increase. The estimated annual cost curve of the new vehicle is superimposed on the known and projected old vehicle cost curve to surface the break even point. After this point, savings in increasing amounts will be realized by buying a new vehicle.

Admittedly, the estimation of new truck costs is far from being precise due to the complex pattern of price and productivity changes; however, this method is better than "crystal-balling" the decision.

One of the trends to evaluate closely on new vehicles is the ever increasing cost of fuel and maintenance caused by continuously increasing pollution control equipment. As a further guide, the following table shows frequently used rough guidelines for replacement based on years of service:

- Cars 5 years
- Light trucks 7 years
- Medium trucks 9 years
- Heavy trucks 10 years

PLANNING

General

Acquisition of vehicles when they are needed involves three elements:

- Forecasting of requirements
- Estimating of process flow times:
  - Time to forecast
  - Paper processing and approval time
  - Purchasing and delivery time
  - Retrofitting time
- Expediting throughout the processing system

A discussion of these elements follows:

Forecasting of requirements

Forecasting of manpower and equipment requirements utilizing the principles of extrapolative and entrepreneurial planning are fundamental to the success of a vehicle acquisition program. Each department head is responsible for planning vehicle requirements sufficiently in advance to allow for the varying processing times of each vehicle type.

Process Flow Times

The garage superintendent shall be responsible for

publishing on an annual basis the processing time for each different vehicle type.

Expediting

Expediting is frequently a major problem in on-time delivery since each different person involved in each processing step has a tendency to "drop the ball" upon completion of his portion of the overall process. Each person involved in the various processing steps defined timewise in the paragraph above shall be responsible for completing the transaction within the prescribed time allotted.

SELECTION OF EQUIPMENT

General

There are several aspects to the problem of selecting new or replacement equipment items, these are:

- o How to determine correct type and quantity.
- o Selecting the best available on the market to match type selected.
- o The need for standardization.
- o The special problem of excess make/type combinations.

These items are discussed in the following paragraphs.

Selecting the Type and Quantity

Each department performs various types of jobs. Each job type represents a certain percentage of the total work effort. For example, let us assume that the installing of a sewer line is one job type, and that this job type represents 25% of the total department labor effort. Let us further assume that an analysis has been made and a determination made that vehicle type "X" is most suitable for this type of work. It, therefore, would follow that 25% of the labor force should be equipped with vehicle type "X." This reasoning process is complicated by the fact that requirements, loads, and job mix percentages are constantly changing.

The point to be made by the above discussion is that in order to make an intelligent choice of new equipment type and quantity, someone must:

- o Categorize jobs by type.
- o Determine workload percentage distribution by job type.
- o Select proper equipment for each job type and crew size and mix.
- o Monitor equipment acquisition to conform with the above items.

It shall be the responsibility of each department head to maintain this information for backup justification data on all new and replacement vehicle acquisitions.

Selecting the Best Available Equipment

Once the equipment type and quantity are determined, someone must match what's available on the open market with this requirement. This shall be the responsibility of the garage superintendent who must be the plant's expert in this regard. He shall maintain this expertise by constantly monitoring publications, advertisements, and product exhibitions.

When a new product is advertised that can do an existing task better, he must notify the affected department head, so that he may consider it during the next budget preparation.

Standardization of Specifications

The garage superintendent shall be responsible for developing and maintaining a standard set of specifications for frequently purchased equipment types.

The purpose of these specifications are:

- o To facilitate competitive bidding.

- To keep the number of different vehicle types to a minimum.
- To insure that everything that is needed is specified. (Each standard specification on a vehicle frequently purchased will improve with age as mistakes and omissions are corrected.)

Specifications proposed on the spur of the moment frequently omit important physical items or performance criteria. Standard specs help reduce these problems.

The Problem of Excess Make/Type Combinations

The problem of excessive make type combinations is illustrated in Figure 5-14. (A sample list from a public utility is used to illustrate the problem.)

This figure shows that there are 110 different make/-type combinations which is unnecessarily high! Too many different types will cause the following hardships:

- Excessive parts procurement and stocking problems.
- Excessive maintenance skill requirement.
- Since each vehicle type generates its own type of repetitive failures, the maintenance analysis problem is compounded.
- Makes more complex the writing of preventive maintenance instructions.

Typical reasons for having too many vehicle types are:

- Purchasing policy of securing competitive bids and giving the award to the lowest bidder.
- Desire of the plant to buy from several local vendors for goodwill purposes and to facilitate warranty work.
- The pressure to use the latest and best vehicle for the job for methods improvement reasons.

Make	Number of Types	Number of Vehicles	Cost In $	Percent By Number	By Cost
Invalid	1	15	365,704	1.0	2.8
Buick	1	2	12,926	0.1	0.1
Chevrolet	15	548	2,071,868	37.3	15.8
Dodge	9	29	128,924	2.0	1.0
Ford	28	411	2,314,356	28.0	17.7
FWD	7	28	995,257	1.9	7.6
GMC	4	12	49,598	0.8	0.4
International	28	299	6,532,038	20.3	50.0
Jeep	3	3	12,393	0.2	0.1
Mack	2	2	13,741	0.1	0.1
Plymouth	4	97	308,400	6.6	2.4
Rambler	1	2	6,064	0.1	0.0
Pontiac	2	14	48,094	1.0	0.4
FWD	2	2	111,402	0.1	0.9
GMC	1	1	3,994	0.1	0.0
Cushman	1	4	6,769	0.3	0.1
Hydrol	1	1	91,204	0.1	0.7
TOTALS	110	1470	13,072,732	100.0	100.0

Figure 5-14: Distribution of vehicles by type and investment.

All of the above items are valid and good ground rules to follow *providing* they are:

- o Used with restraint.
- o Counterbalanced by someone who has the disadvantages of excessive vehicle types in mind and who will use his ingenuity to stay within the ground rules and, at the same time, keep vehicle types to the minimum.

The garage superintendent shall spearhead a drive to reduce the numbers of vehicle types. A periodic tabulation similar to Figure 5-14 shall be made to follow up on the progress of this reduction effort.

DISPOSAL

A policy should be established concerning the disposal of surplus fleet equipment. Some ideas that may be considered are described below:

Consideration may be given to auction off, or sell directly, the equipment to employees of the company. Sale price could be the red book value.

Surplus and obsolete materials and equipment should be reported to the purchasing department, and disposition be made as follows:

- Transfer to other departments. The purchasing agent, upon approval of the director of finance, will have the authority to transfer surplus stock to other using department.
- The sale of equipment and other assets will be authorized by the plant engineering manager. Upon approval by the plant engineering manager, the purchasing department will advertise the goods for sale to the highest bidder. It will not be necessary to formally advertise the sale of extremely small equipment.
- Trade-in on new equipment. The using department may trade in old equipment on new equipment.
- Junking. All materials or equipment that is junked may be disposed of by the purchasing agent. In the case of rolling equipment, however, the plant engineering manager must approve its removal from the equipment list. Once a piece of equipment is no longer serviceable, it

should be reported immediately so that it may be removed from the equipment list and the insurance company can be notified for return of part of the premium.

LEASE VERSUS BUY CONSIDERATIONS

Because plants at times may have available low interest rates on borrowed money, compared with rates that must be paid by leasing companies, it may be advantageous for the plant to own its own vehicles. This is not the case, however, for vehicles used intermittently or seasonally.

There are two management information systems that should be utilized in helping each department head decide on whether or not to buy or lease equipment which is to be used part time, these are:

- The miles or hours per month per vehicle statistics published as a part of the Transportation Management Information System. (See Chapter 29).
- The dispatch control system for intermittently used vehicles described previously.

These information systems will surface the nature and extent of the intermittent use. The transportation information system will provide ownership cost data on similar units for comparison with leased rates. This information will enable the department head to make the logical decision on whether to buy or lease.

29

Transportation Management Information System (TMIS)

GENERAL

The TMIS shall be a monthly computerized report segregated by vehicle number. This report is very useful because it has practically all of the information that you might want on one line. This makes the information very compact and easily accessible since you can scan each vehicle number back and forth a line or a page at a time. Approximately 50 vehicles can be put on one page, therefore if you have 200 vehicles the report will only take up 4 pages. In addition this system combined with the computerized vehicle PM system will allow the plant to utilize operating hours or mileage instead of elapsed time for monitoring the maintenance program not only on vehicles but on any equipment that is more logical to base PM on hours or miles. The format of this report is shown in Figure 5-15. Input for the TMIS shall consist of the following documents:

- o Garage Work Order
- o Preventive Maintenance Order (Labor)
- o Garage Service Work Order (Labor)
- o Invoices (Material)
- o Service Ticket (Material)
- o Fuel Ticket (Material)
- o Storehouse Material Requisition

1	2	3	4	5	6	7	8	11	12	13	14	15	16	17	18	19	20	25	26	27	28	29
VEHICLE NUMBER	PURCHASE DATE	PM $ M	BM $ M	OM $ M	BM_H M	D_H M	MILES M	PM $ S	BM $ S	OM $ S	TOT $ M	TOT $ S	MILES S	BM/TM M	BM/TM S	DT/BH M	DT/BH S	MPG M	MPG S	CPM M	CPM S	MPY
XXXXXXXXXXX	XXXXXXXX	XXXXX	XXXXX	XXXXX	XXX	XXX	XXXX	XXXXXX	XXXXXX	XXXXXX	XXXXX	XXXXXX	XXXXXX	XX	XX	XXXX	XXXX	XX	XX	XXXX	XXXX	XXXXX

SEE FIGURE 5-18
FOR EXPANDED
DEFINITION OF
COLUMNS.

S=SINCE ACQUISITION
M=MONTHLY

Figure 5-15: TMIS report, report date June 1988.

The following paragraphs describe the processing steps to be followed for each of the above input documents.

THE GARAGE WORK ORDER (LABOR)

An explanation of how to fill out and check the garage work order is given in a previous paragraph. The following is a list of the processing steps after the form is completed, checked and signed:

No.	Description	Responsibility
1	Transmit original to computer clerk	Garage clerk
2	Input to computer	Computer clerk
3	Transmit original to garage clerk	Computer clerk
4	File original	Garage clerk

PREVENTIVE MAINTENANCE ORDER (LABOR)

An explanation of how to fill out the preventive maintenance order is given in a previous paragraph. The processing of the PM order is identical to that described for the garage work order in the above paragraph.

SERVICE WORK ORDER (LABOR)

Service labor includes tasks such as:

- Pumping gasoline
- Adding oil
- Washing vehicle
- Road calls for service work
- Greasing
- Pickup of vehicles
- Steam cleaning

- o Wheel packing
- o Wheel balancing
- o Tire rotation

The labor charge for each of these items per vehicle is quite small. The cost of accounting for these charges would be too great a percentage of the job cost. For this reason, it is recommended that this not be done. Instead, service labor charges should be lumped together and redistributed over each vehicle on the basis of its gasoline consumption, for example:

$$\text{Service Charge for Vehicle "X"} = \text{Total Service Charge for All Vehicles} \times \frac{\text{Gasoline Consumption for Vehicle "X"}}{\text{Total Gasoline Consumption}}$$

This calculation is a reasonable approximation of the actual service labor charge. Considering the fact that the total service *labor* cost is a relatively small portion of the overall annual operating expense, any variation between actual and estimated costs will be insignificant.

An explanation of the mechanics of calculating the *total* service labor charge follows. The garage clerk shall post to a "Garage Service Work Detail Sheet." (See Figure 5-16.) One of these sheets is required for each serviceman. Servicemen should perform two types of work, i.e.:

- o Service Work
- o Preventive Maintenance

They may also be involved in repair work if they run out of service work. Whatever they spend on PM and repair work shall be recorded on Figure 5-16. The difference between the *total* hours worked that day and the hours worked on PM and breakdown repairs shall be recorded as service hours. A running total shall be calculated each day and placed in the cumulative date box.

Month ____________ Serviceman's Name ________________ Rate _____

Day of Month	PM Hours	Breakdown Repair Hours	Service Work Hours	Service Work Hours Cumulative To Date
26				
27				
28				
29				
30				
31				
1				
2				
3				
4				
5				
6				
7				
8				
9				
10				
11				
12				
13				
14				
15				
16				
17				
18				
19				
20				
21				
22				
23				
24				
25				

Total for Month __________ Hrs.

______ Rate X ______ Total Hours = ______ Total Dollars for Month

Figure 5-16: Garage service work detail sheet.

At the end of the month, the clerk shall multiply the total service hours by the serviceman's rate to arrive at the particular serviceman's monthly charge. The monthly charge for all servicemen shall be summarized as shown on Figure 5-17. This calculation shall be checked and approved (signed) by the garage superintendent. A copy of the form shall be retained and filed by the garage clerk. The garage clerk shall then transmit the original to the computer clerk. The computer clerk shall program the computer to calculate the service charge for each vehicle in accordance with the previously explained equation. All service labor shall be charged to an "Operation" account.

SUMMARY OF MONTHLY GARAGE SERVICE CHARGES

for the month of __________, 1988

	Serviceman's Name	$ Service Charge
1.		
2.		
3.		
4.		
5.		
6.		
7.		
8.		
9.		
10.		
11.		
12.		
	Total $ Charge	__________

CHECKED: ______________________
Garage Superintendent

Figure 5-17: Summary of monthly garage service charges.

ALL MATERIAL CHARGES

Material charges should be picked up from the following documents:

- o Invoices
- o Service Tickets
- o Fuel Tickets
- o Warehouse Material Requisitions

COMPUTER FUNCTIONS

All input documents to the TMIS should be sent to the computer clerk. This clerk will have the responsibility of processing the input data through the computer to produce the monthly TMIS. Figure 5-18 describes the database fields and how the fields are obtained. Column numbers used in this figure refer to column numbers used in Figure 5-15. Notice that column numbers 9, 10 and 21 through 24 have been left out of the report even though they are included in the data base. The reason for this is that these columns are used primarily to calculate other columns and are not of much use otherwise.

SEQUENCE OF TMIS MONTHLY TAB RUN

The TMIS monthly tab run shall be listed in numerical order of the 12 digit vehicle number. This will list the vehicles by department and vehicle type.

DEFINITION OF "SINCE ACQUISITION"

Values in the TMIS labeled "S" for since acquisition refer to costs accumulated since the vehicle purchase date. However, since the report will start with zero cumulative data, the columns labeled "S" for since acquisition will actually mean "since report start date." This fact must be kept in mind when analyzing the data unless you want to go through the trouble of putting in the old data for the vehicle existing before the TMIS system was installed.

COL	DESCRIPTION	NAME	NO. DIGITS	COMMENTS
1	VEHICLE NO.	VEH_NO	12	DOESN'T CHANGE
2	PURCHASE DATE	PURCH_DA	8	DOESN'T CHANGE
3	PM COST FOR MONTH	PMCOST	5	INPUT
4	BM COST FOR MONTH	BMCOST	5	INPUT
5	OM COST FOR MONTH	OMCOST	5	INPUT
6	BM HOURS FOR MONTH	BMHRS	3	INPUT
7	DOWNTIME HRS FOR MONTH	DOHRS	3	INPUT
8	MILES/HRS FOR MONTH	MLSMO	4	INPUT
9	GALS FOR MONTH	GALSFUEL	3	INPUT
10	DATE OF THIS REPORT	DAYRPT	8	INPUT
11	PM COST SINCE ACQUIS.	PMCOST_S	6	LAST MO. + COL 3
12	BM COST SINCE ACQUIS.	BMCOST_S	6	LAST MO. + COL 4
13	OM COST SINCE ACQUIS.	OMCOST_S	6	LAST MO. + COL 5
14	TOTAL MONTHLY COSTS	TMCOST	5	COL 3 + COL 4 + COL 5
15	TOTAL COST SINCE ACQUIS.	TC_S	6	COL 11+COL12+COL13
16	MILES/HRS SINCE ACQUIS.	MILES_S	6	LAST MO. + COL 9
17	% BM/TM COST FOR MONTH	%BM/TM/M	2	COL 4 ÷ COL 14
18	% BM/TM COST SINCE ACQUIS.	%BM/TM/S	2	COL 12 ÷ COL 15
19	RATIO DT/BH FOR MONTH	DT/BH/M	4	COL 7 ÷ COL 6
20	RATIO DT/BH SINCE ACQUIS.	DT/BH/S	4	COL 24 ÷ COL 23
21	MONTHS IN SERVICE	MIS	4	CALC. FROM COL 2
22	GALLONS SINCE ACQUIS.	GSA	5	LAST MO. + COL 9
23	BM HOURS SINCE ACQUIS.	BMS	5	LAST MO. + COL 6
24	DOWNTIME HRS. SINCE ACQUIS.	DHS	5	LAST MO. + COL 7
25	MILES/HRS PER GAL. FOR MO.	MPG_M	2	COL 8 ÷ COL 9
26	MLS/HRS PER GAL SINCE ACQ.	MPG_S	2	COL 16 ÷ COL 22
27	CENTS PER MILE FOR MONTH	CPM_M	4	COL 14 ÷ COL 8
28	CENTS PER MILE SINCE ACQ.	CPM_S	4	COL 15 ÷ COL 16
29	MILES/HRS PER YEAR	MPY	5	MILES FOR PAST YEAR

Figure 5-18: List of fields–TMIS report.

Eventually, as vehicles are retired and new vehicles purchased, the system will automatically purge itself and since acquisition will acquire its true meaning.

ANNUAL SUMMARY

Once per year, on the first day of the year, the computer should be programmed to print out an annual summary as described below. For each vehicle the total cost shown in column 15 should be subtracted from the total cost of the previous year on the same date to produce the annual total cost. This figure shall be printed by vehicle in the same sorting sequence previously described. The computer shall then show the average annual cost by department and type and the average cost by type for all departments. This information will be utilized to surface deviations from average and for replacement analysis purposes. The format of the report shall be as follows:

Vehicle Number	Months in Service	Annual Operating Cost

30

Use of the TMIS

GENERAL

The primary use of the TMIS will be obtained from columns 17-20 and 25-29. Standards for each of these parameters should be set for each vehicle type by the garage superintendent with the help of the department head. The setting of standards can be performed by using one or more of the following techniques:

- o Published standards
- o Personal experience
- o Using the average value of the best performers

Standards are in effect a goal to shoot at and a reference point about which to measure deviations.

PERCENT BREAKDOWN TO TOTAL MAINTENANCE COST

This figure is a measure of the success of the PM program. It is suggested that an initial target goal of 15% be set.

RATIO OF DOWNTIME TO BREAKDOWN MAINTENANCE COSTS

Column 12 shows the monthly downtime hours which in

itself shows if there is a downtime problem. The ratio of downtime to breakdown maintenance hours surfaces a different kind of problem. To illustrate this, let us assume that a ladder truck in the fire department broke down from a faulty ignition coil. Since the fire department has only one ladder truck and since the criticality code would be "A," it is assumed that the garage would give the repair top priority. If the truck was down for two hours and the repair time was two hours, the ratio of downtime to repair time would be one and this would be good. However, suppose the truck was down for two weeks. The ratio would then be:

$$\frac{2 \times 7 \times 24}{2} = 168$$

This ratio would surface some type of problem. The unfavorable ratio could have been caused by:

- o Material acquisition time
- o Labor shortage
- o Arbitrary decision not to work on the truck due to higher priority work

Whatever the reason, this ratio will surface the problem if it arises.

MILES OR HOURS PER GALLON

Deviation from the standard set for this item will surface a problem with engine efficiency.

COST IN CENTS PER MILE/HOUR

This figure is an overall barometer of the economy of operation.

MILES OR HOURS PER MONTH SINCE ACQUISITION

This figure is a measure of the vehicle utilization and will surface inadequately utilized vehicles. This will then cause a special study to be made as described in Chapter 27.

Appendix

Cost Effective Maintenance Management Computer Program

INTRODUCTION

General

The maintenance management program contains six database systems, the work order database, the preventive maintenance database, the equipment inventory record database, the fleet equipment PM, the fleet management control and the operating hour/mileage databases. All systems are written in Dbase III (Dbase III is a trademark of Ashton-Tate) and are compiled with Clipper (Nantucket Software) and are run without requiring Dbase III software. The work order database is used to record and report the status of work orders, to facilitate failure analysis and to facilitate maintenance optimization. The preventive maintenance (PM) database contains the PM program and provides for input and output to the PM system. The equipment record system is used to record nameplate and other data and to list spare parts kept for the equipment. The fleet equipment databases are written specifically for fleet equipment and cater to the differences in that type of equipment. All databases are accessed through master and sub-master menus and are menu-driven and hence are user-friendly requiring practically no computer experience or knowledge.

System Requirements

All databases are written for an IBM PC/XT/AT

computer or compatible system with one or more floppy disks and a hard disk with at least 10MB capacity. Internal memory capacity required is 640K and an internal clock is mandatory.

GETTING STARTED

Symbols and Notes

- R means to hit the return (enter) key.
- Only those characters enclosed in brackets [ ] are meant to be typed.
- Spaces shown between characters in brackets are mandatory.
- C> or A> means the "C" or "A" prompt.

Making backup copies

It is recommended that you make backup copies of program disks nos. 1, 2 and 3 and store your originals in a safe place. This can be done as follows:

If you have one floppy drive:

1. Boot up computer with your DOS disk in A drive.
2. At the A>, type [DISKCOPY], R
3. Follow instructions to copy all program disks.

If you have two floppy drives:

1. At the C>, type [DISKCOPY A: B:], R
2. Follow instructions to copy all program disks.

You need three blank 5 1/4" double sided, double density disks to copy onto. They need not be formatted since the diskcopy procedure formats as well as copies.

Copy Program Disks onto Hard Disk

Copy maintenance program disks 1, 2 and 3 into a separate directory as follows:

At C> in root directory:

1. Type [MD/MAINT], R
2. Type [CD/MAINT], R
3. Put maintenance program disk no. 1 in A drive
4. Type [COPY A:*.*], R
5. Repeat steps 3-4 for maintenance program disks 2 and 3.

This will copy the maintenance program onto your hard disk in a separate directory that you have named "MAINT."

To Start Maintenance Program

1. Change to "MAINT" directory by typing [cd:\MAINT], R
2. At C> type [MAINT], R
3. This will bring up a master menu that looks like this:

MASTER MENU

****SELECT DATABASE, TYPE ITS NUMBER THEN HIT THE RETURN KEY****

1: WORK ORDER

2: PREVENTIVE MAINTENANCE

3: EQUIPMENT INVENTORY

4: FLEET EQUIPMENT PM

5: FLEET MANAGEMENT CONTROL

6: OPERATING HOUR/MILEAGE REPORT

---

4. Just follow the instructions to select whichever database you want. When you exit from each of the databases, you will

be returned to the C>. Then you can either select a different database or change to a different directory.

5. See instructions at the end of this guide for using the demonstration (demo) disks.

Miscellaneous

For consistency through the database, set the CAPS LOCK on.

The hard disk containing the data and the software should be backed up at frequent intervals in case you should inadvertently lose the data on the hard disk.

THE WORK ORDER DATABASE

General

If you selected item 1 in the master menu, you will call up the work order database and see the following menu:

A: Create a new Work Order

B: Print a Work Order Form

C: Edit an Existing Work Order

D: Work Order Run by Machine Number

E: Work Order Run by Requesting Cost Center

F: Work Order Run by Performing Cost Center

G: Work Order Run by Work Type Code

H: Search by Machine Criticality Code

I: Search by Machine Number

J: Failure Analysis Report

K: Annual Machine Maintenance Cost Analysis

L: Delete a Work Order

M: Brief Summary of Work Orders

N: Backlog Report

O: Exit

Simply select the menu option you desire and hit that letter and then follow instructions.

Creating a new work order takes place in three sessions. First, opening information is entered under option A: "Create a New Work Order." Secondly, the planning and estimating information is entered under option C: "Edit an existing work order." This part may be combined with the first step if desired, particularly if the job is small and simple. Thirdly, the remainder of the information is added when the work order is complete under option C: "Edit an Existing Work Order."

To Open a Work Order

The following items should be entered in the work order record to open the WO:

- Work Order Number*
- Machine Number*
- Machine Description
- Work Requested
- Requesting Cost Center Number
- Performing Cost Center Number
- Requestor
- Request Date
- Priority
- Work Type
- Job Title

An explanation of all of these items is contained in the textbook . Comments on the two starred items are given below:

Work Order Number

There are 10 digits for your use in designing a WO number. The design should suit each plant's needs. However, a suggested design is shown below:

Digits 1&2	-	Month
Digits 3&4	-	Year
Digits 5&6	-	Area code (divide plant into 0-99 areas)
Digits 7-10	-	Sequence no. (0 to 9999)

Machine Number

There are 12 digits for your use in designing a machine number which should be designed to fit your plant's needs. A recommended design is shown below:

Digit 1	-	Criticality code
Digits 2&3	-	Equipment category
Digits 4&5	-	Equipment subcategory 1
Digits 6-10	-	Sequence no. (0 to 9999)
Digits 11-12	-	Reserved for PM instructions

The criticality code is necessary to enable sorting by this code, a very important convenience.

An example of the equipment categorization is:

Equipment category AIR CONDITIONING SYSTEM

Equipment subcategory 1 COMPRESSOR (a component of the air conditioning system)

The equipment categorization codes will enable sorting by equipment types, a worthwhile tool. However, if you are not interested in this convenience you can simply use the 12 digit number as a sequence number.

Planning and Estimating Stage

After planning and estimating are complete, the following items on the work order are filled by using C: "To edit an existing WO":

- o Estimated crew size
- o Estimated crew hours
- o Estimated material cost
- o Estimated labor cost
- o Estimated total cost
- o Approved by:

The purpose of this information is threefold:

- o To provide an estimate of the total labor and material costs to provide a basis for approval.
- o To facilitate scheduling.
- o To provide written approval to proceed.

Information to Add at Work Order Completion

After the work order is complete, the following items on the work order are added using C: "To edit an existing WO"

- o Job title*
- o Work performed
- o Failure code
- o Component name*
- o Actual material cost*
- o Actual labor crew size and hours*
- o Actual labor cost*
- o Actual total cost*
- o Approved by:*
- o Completion date:

These items are discussed in detail in the text. The following additional comments are made concerning the starred items above:

Job Title

If the job title entered during the opening of the WO does not describe the work adequately, it should be changed at this point.

Component Name

Write out the component name that correlates with the failure code.

Actual Material, Labor and Total Costs

Replace the estimated costs with the actual. If there is a difference between the two, it should be addressed at this time rather than to wait to ponder why there is a difference several days or months later when you have forgotten all about the job. If you want to make a detailed investigation of the differences, then the two values can be written side by side on a hard copy.

Approved By

Add name of man approving completion next to the name approving job creation.

WORK ORDER APPLICATIONS

We have shown how to use menu choices A and C and shall now explain the use of the remainder of choices.

B: Print a Work Order Form

Used to print a copy of the work order form.

D: Work Order Run by Machine Number

This run shows all work orders written against a specific machine and is used when analyzing the maintenance history on a specific machine.

<u>E: Work Order Run by Requesting Cost Center</u>

Used principally by the requesting cost center to determine maintenance charges against their cost center.

<u>F: Work Order Run by Performing Cost Center</u>

Used by maintenance manager to determine the spread of workload by performing cost centers.

<u>G: Work Order Run by Work Type</u>

This is a very important and useful tool which is explained in case study 2B.

<u>H: Search by Machine Criticality Code</u>

Displays or prints WO's by criticality code. Facilitates analysis of the most critical machines for the purpose of reducing failures or to create or revise PM instructions on machines having a specific criticality number.

<u>I: Search by Machine Number</u>

Displays or prints all WO's for a specific number.

<u>J: Failure Analysis Report</u>

This is one of the two most important reports in this program. It greatly facilitates the maintenance failure analysis task. The report and the subject is treated in depth in the entire Part 2 "The Downtime Improvement Cycle".

<u>K: Annual Machine Maintenance Cost Analysis</u>

This is the second most important report. It shows in descending order of cost by machine, the percent each is of the total cost and cumulative percentages. The costs are 12 months moving cumulative maintenance costs for each machine. (This is why an internal clock is required in your computer so that the computer can read a year's worth of information starting from the current date, which changes constantly.) This permits quick analysis of where the big costs are to allow concentration on the biggest problem. It uses the principle that the majority of costs lie in a few machines, whereas most of the machines normally account for a small portion of the costs. This report is so powerful in what it can do that I call it "The Magic Tool," for that is what it is. It is the key tool to be used as explained in Part 1, "Maintenance Optimization."

L: Delete a Work Order

This option will prompt for a 10 digit work order number which the system will delete from the database file.

M: Brief Summary Report of Work Orders

The purpose of this option is to provide a single line report for each work order. The single line contains the WO number, machine number, request date, completion date and the job title.

This will allow scanning of all work orders in much shorter time and facilitate looking up WO's and machine numbers. You have the choice of sorting this report three ways:

1. By work order number
2. By machine number
3. By request date

You also have the option of sending the report to the printer or not. If the report is sent to the printer, all work orders are printed out using one line per WO. If you elect no (N), then the report is sent to an ASCII file with a title "WOBRIEF.TXT." In order to call this report up on the screen you must:

1. Go back to the C>MAINT.
2. Type [list WOBRIEF.TXT], return

You can then scroll through the report line or page at a time, backwards, forwards or sideways by using the cursor keys (2, 4, 6 or 8) or the page up or page down keys.

Everytime you call up this file (WOBRIEF.TXT), the computer will update the file with the latest information.

N: Backlog Report

The purpose of this report is to provide a single line backlog of work report. This report contains the request date, WO number, machine number, job title and estimated hours broken down by trade. Trade hours of backlog are also totaled. This report can be manipulated in the same manner as the "M" report. The file name in this case is "BACKLOG.TXT."

O: Exit

Provides an exit from the program.

THE PREVENTIVE MAINTENANCE DATABASE

General

If you selected item 2 in the master menu, you will call up the Preventive Maintenance database and see the following menu:

A: Add new PM instructions

B: Edit existing PM instructions

C: Weekly PM task schedule

D: Annual PM costs by machine number

E: Annual PM cost analysis

F: Print all PM instructions

G: Delete an instruction

H: Preventive Maintenance Report (Brief)

I: Exit

Again, simply select the menu option you desire and hit that letter and then follow instructions.

The user of the PM database should read the entire Chapter 13 entitled "Preventive Maintenance".

The PM database specifically follows the system entitled "Preventive Maintenance Instructions (PMI's)". The instructions that follow supplement the instructions in the book and explain some minor changes that have been made to the system described in the book to facilitate computer programming.

Entering New PMI

First, open the PM database as described above. Next, select menu item A (add new PM instructions). The PMI computer input form (Figure 2-16) must be available for the new PMI's to be entered. A separate instruction must be entered for each unique item on the PMI. For example, item no. 4 (renew belts) is filled out below. Notice that all items are for all practical purposes self explanatory.

FOREMAN NUMBER	:	1
MACHINE NUMBER	:	682958000000
MACHINE DESCRIPTION	:	FAN
TRADE: E, L, M	:	MECHANICAL
INSTRUCTIONS	:	RENEW BELTS
INSTRUCTIONS PART II	:	
ESTIMATED HOURS	:	.3
ESTIMATED COST:	:	20.00
WEEK CODE	:	
MONTH CODE	:	
QUARTERLY CODE	:	
SEMI-ANNUAL CODE	:	
ANNUAL CODE	:	52
BI-WEEKLY CODE	:	

It is important to note, by reading the textbook, that the scheduling code shown above provides for automatic scheduling of each task and for spreading the workload as evenly as possible throughout the year.

The last two digits of the machine number are reserved

to distinguish between different instructions for the same machine number.

The method of selecting the scheduling code is explained in the preventive maintenance section of the textbook, however, notice above that a bi-weekly code has been added to the program that does not show in the textbook.

The bi-weekly code should be given the number 97 or 98 depending on whether you want these tasks to start on the first or second week of the month. (See the scheduling code table). Also the bi-weekly task column should be added to the PMI computer input form.

Edit a PMI

Select letter B, to edit an existing PMI, then follow instructions.

To Print Out Weekly PM Schedule

Select menu item C, to print out the weekly schedule. The printout will be sorted by foreman, machine number and trade to facilitate handing out instructions by foreman, machine number and trade. Follow instructions in the textbook for proper processing of the weekly PMI's.

Reports

The PM software provides for two types of reports:

- o Annual PM costs by machine number (menu selection D)
- o Annual PM cost analysis. This is an annual cost by machine number in descending order to match the report of breakdown costs in the work order database. (Menu selection E)

These reports provide knowledge of total machine costs by machine number.

In addition there is the (H) report which contains on one line the machine number, machine description, foreman number, estimated hours per trade totals and cost totals. Report is sorted by machine number. This report can be manipulated the same way as the other reports. The file name is "PMBRIEF.TXT."

Print All PM Instructions

MENU SELECTION F prints all PM instructions in the system, ordered by machine number.

Delete Instructions

MENU SELECTION G will delete a set of PM instructions for a given machine number.

To Exit Program

Select (H) to exit from menu. The system will return to the C>.

THE EQUIPMENT INVENTORY RECORD DATABASE

Introduction

The purpose of the equipment inventory record system is to record vital information about each item of equipment and the spare parts maintained for the equipment.

If you selected item 3 in the master menu, you will call up the equipment inventory record system and see the following menu:

```
------------------------------------------------------------
        EQUIPMENT DATABASE - MENU SELECTION

    A:  CREATE A NEW EQUIPMENT RECORD
    B:  PRINT AN EQUIPMENT RECORD FORM
    C:  EDIT AN EXISTING EQUIPMENT RECORD
    D:  DELETE AN EXISTING EQUIPMENT RECORD
    E:  REPORT BY MACHINE NUMBER
    F:  REPORT BY LOCATION
    G:  REPORT BY DEPARTMENT NUMBER
    H:  EXIT

ENTER YOUR CHOICE:
------------------------------------------------------------
```

Again, simply select the menu option you desire and hit that letter and then follow instructions. The function of each selection is:

A: Create a New Equipment Record

The purpose of this selection is to create a new record. Simply type in the first item and hit return which will automatically advance you to the next item. When you finish the last item, hit return and a screen will come up to allow you to enter the spare parts for the machine. You can use any format you want for the spare parts, however, a suggestion is shown below:

Stock #	Mfg's #	Description	Min/Max	Unit Cost $

B: Print an Equipment Record Form

Used to print a specific equipment record.

C: Edit an Existing Equipment Record

Used to edit an existing record.

D: Delete an Existing Equipment Record

Used to delete an existing record.

E: Report by Machine Number

Creates a list of all machines sorted numerically by machine number with one line per machine. Manipulate same as other one line reports. Report ASCII name is EQ1.TXT.

F: Report by Location

Creates a list of all machines sorted numerically by location. Manipulate per other one line reports. Name is EQ2.TXT.

G: Report by Department Number

Creates a list of all machines sorted numerically by department number. Manipulate per other one line reports. Name is EQ3.TXT.

H: Exit

Used to exit to master menu.

One important point to be aware of is that lists (options E, F, and G) under program 3 (EQUIPMENT RECORDS) are meant to be printed on wide paper. If you use 8 1/2 x 11 paper you can get everything on by using condensed print (15 CPS).

FLEET EQUIPMENT PM DATABASE

General

If you selected item 4 in the master menu you will call up the fleet preventive maintenance database and the following menu:

FLEET PREVENTIVE MAINTENANCE DATABASE
MENU SELECTION

CHOICE:

A: ADD NEW PM FLEET INSTRUCTIONS

B: EDIT EXISTING PM FLEET INSTRUCTIONS

C: WEEKLY PM FLEET TASK SCHEDULE

D: PM FLEET COSTS BY MACHINE NUMBER SINCE ACQUISITION

E: PM FLEET COST ANALYSIS SINCE ACQUISITION

F: PRINT ALL PM FLEET INSTRUCTIONS

G: DELETE AN INSTRUCTION

H: FLEET PREVENTIVE MAINTENANCE REPORT (BRIEF)

ENTER YOUR CHOICE:

Simply select the menu choice you desire and hit that key.

Entering a New Fleet PM Instruction

Select choice A. An input screen will appear. Fill out the various items which are essentially self-explanatory.

Notice that the form allows only 5 PM instructions to be added. An additional 130 instructions my be added by reserving the last digit of the machine number to specify sets of 5 instruction items. A letter (i.e., A, B, C, etc.) can be used to differentiate the sets.

The current miles/hours should be updated periodically (say, weekly) in the edit mode. (Option B)

Notice that either miles or hours may be used as a basis for kicking out the PM instruction automatically. The "miles" column just to the left of the "instruction" column is where the frequency of PM is inserted. The foreman column provides a place to segregate PM instructions by foreman. However this column can be used to specify different plants, buildings or locations instead of foremen.

Edit an Instruction

Choice B, provides the facility to modify instructions. This is where you should update the mileage/hours every week.

Weekly PM Fleet Task Schedule

Choice C produces the weekly PM schedule for those equipment items due by the computer comparing the chosen frequency of miles or hours with the current mileage or hours.

When you hit "C," the screen will ask you to enter the mileage/hour tolerance. If you enter 100, any current mileage within plus or minus 100 miles of the frequency mileage will be selected. Assume you specified an oil change every 1,000 miles. A current mileage of 3,950 would be selected because it is within the tolerance and it is a multiple of 1,000, whereas 3,850 would not be selected.

Menu Options D Through G

Option D provides a summary of cost sequentially by equipment item, since acquisition.

Option E shows the same thing as D except that it is arranged in descending order of cost to facilitate maintenance optimization.

Options F and G are self-explanatory.

Brief Report

Option H provides a brief report that puts critical information on one line. This report can be sent to the printer or it can be called up on the screen.

In the latter case, you can scroll back and forth, line or page at a time or left to right. Thus, this feature can be used to quickly locate an instruction.

Notice that if you select the option of not sending the report to the printer, a message appears saying that the report is written to file: "PMBRIEF.TXT." If you press any key, the report will scroll by. Then the main menu appears. Select exit, option I. This will bring you back to a C> in the directory where the program is installed. Then type LIST PMBRIEF.TXT. This will bring up the report on the screen which will allow you to scroll back and forth.

FLEET MANAGEMENT CONTROL (FMC) DATABASE

General

This database provides "all the information you wanted to know about your fleet but were afraid to ask." There are 28 columns of information on one line consisting of 10 input columns and 18 columns that are calculated from the input columns. Having all the information on one line makes the report very compact. If you have 200 vehicles, the entire report can be printed on 4 pages. In order to permit printing on one line, wide paper and compressed print capability (15 cpi or higher) are required.

FMC Menu

Choosing option 5 on the master menu brings up the FMC menu which is:

FLEET MANAGEMENT CONTROL SYSTEM SELECTION

CHOICE:

A: **ADD NEW RECORDS**

B: **EDIT EXISTING RECORDS**

C: **PRINT RECORDS**

D: **DELETE RECORDS**

E: **FMCS REPORT**

Includes ability to scroll through report using the LIST.COM function.

F: **CALCULATE MODE**

When the user hits F return, the program automatically calculates all of the new information just entered.

G: **EXIT**

An explanation of each option follows:

Option A: Add New Records

Use this option when adding a new machine for the first time. Always follow up this step by utilizing option F "The Calculate Mode." Please note that once you enter a specific machine number, option A can no longer be used for that machine number. From then on option B, "Edit Existing Records" must be used to update that machine.

Option B: Edit Existing Records

Use this option to periodically update machine records.

A weekly update is suggested. A monthly update should be the minimum update used.

Option C: Print Records

This option prints all machine records in long form and includes all 28 columns of information. The headings are easier to read than in option E.

Option D: Delete Records

This option permits you to delete a machine from the database.

Option E: FMCS Report

If you select this option, the following message will appear on the screen:

REPORT OF ALL FLEET MANAGEMENT CONTROL SYSTEM

Report to File FMCS.TXT (otherwise to Printer) Y or N

If you select N (No), then the report will be printed out. Have your printer on and wide paper installed. The computer will automatically specify condensed print (15 cpi). This printout has the wonderful feature of putting all the information for one machine on one line which produces a very compact report. In addition, by careful design of the machine number, equipment of like type will appear together allowing comparison of performance of like type vehicles. The report also permits easy comparison of performance against established performance standards, i.e., 20 miles per gallon plus or minus 5%.

If you select Y (Yes), then the information will be printed to a file named "FMCS.TXT." To read this file on the screen do the following:

1. Hit Y, then return.
2. Select Option G to exit.
3. At the C>, type [LIST FMCS.TXT], return.

This will put the report on the screen and allow you to

scroll up and down, left and right, a line or page at a time.

The headings for this report are drastically condensed to allow them to fit on one line. The condensed headings are clarified in the chart shown below.

FLEET MANAGEMENT CONTROL SYSTEM REPORT
LIST OF FIELDS, ABBREVIATIONS AND FORMULAS

COL	DESCRIPTION	ABBREV.	FORMULA
1	VEHICLE NO.	VEH_NO	DOESN'T CHANGE
2	PURCHASE DATE	PURCH_DA	DOESN'T CHANGE
3	PM COST FOR MONTH	PMCOST	INPUT
4	BM COST FOR MONTH	BMCOST	INPUT
5	OM COST FOR MONTH	OMCOST	INPUT
6	BM HOURS FOR MONTH	BMHRS	INPUT
7	DOWNTIME HRS FOR MONTH	DOHRS	INPUT
8	MILES OR HOURS FOR MONTH	MLSMO	INPUT
9	GALS FOR MONTH	GALSFUEL	INPUT
10	DATE OF THIS REPORT	DAYRPT	INPUT
11	PM COST SINCE ACQUIS.	PMCOST_S	LAST MO + COL3
12	BM COST SINCE ACQUIS.	BMCOST_S	LAST MO + COL4
13	OM COST SINCE ACQUIS.	OMCOST_S	LAST MO + COL5
14	TOTAL MONTHLY COSTS	TMCOST	COL 3+ COL 4+ COL 5
15	TOTAL COST SINCE ACQUIS	TC_S	COL 11+COL 12+COL 13
16	MILES/HRS SINCE ACQUIS	MILES_S	LAST MO + COL8
17	%BM/TM COST FOR MONTH	%BM/TM/M	COL 4 DIV COL 14
18	%BM/TM COST SINCE ACQUIS	%BM/TM/S	COL 12 DIV COL 15
19	RATIO DT/BH FOR MONTH	DT/BH/M	COL 7 DIV COL 6
20	RATIO DT/BH SINCE ACQUIS	DT/BH/S	COL 24 DIV COL 23
21	MONTHS IN SERVICE	MIS	CALC FROM COL 2
22	GALLONS SINCE ACQUIS	GSA	LAST MO + COL 9
23	BM HOURS SINCE ACQUIS	BMS	LAST MO + COL 6
24	DOWNTIME HRS SINCE ACQUIS	DHS	LAST MO + COL 7
25	MILES/HRS PER GAL FOR MO	MPG_M	COL 8 DIV COL 9
26	MILS/HR PER GAL SINCE ACQ	MPG_S	COL 16 DIV COL 22
27	CENTS PER MILE FOR MONTH	CPM_M	COL 14 DIV COL 8
28	CENTS PER MILE SINCE ACQ	CPM_S	COL 15 DIV COL 16

Option F: Calculate Mode

This option *MUST* be used after using either options A or B.

Option G: Exit

This option is used to exit from the program back to the C>.

Copies of reports from choices C and E should be made every month and these should be saved. One use of the saved reports would be, for example, to subtract the values of column 16 a year apart; this would give the annual mileage (or operating hours) on the item of equipment.

OPERATING HOUR REPORT

Purpose

The purpose of this report is to provide a record of the operating hours for any item of equipment. It requires a periodic input of the operating hours (usually daily or weekly) into the computer. When this is done, then an inquiry can be made of the computer to reveal how many hours the equipment has operated between any two dates.

Main Menu

To call up the main menu, type [6], then return. The main menu appears on the screen as follows:

EQUIPMENT OPERATING HOURS DATABASE
MENU SELECTION

CHOICE:

A: ADD NEW RECORDS
B: EDIT EXISTING RECORDS
C: DELETE RECORDS
D: REPORT FOR ALL EQUIPMENT
E: REPORT FOR ONE EQUIPMENT ITEM
F: PRINT ALL EQUIPMENT RECORDS
G: EXIT

The choices are self-explanatory, however, remember to do the following:

- o Keep the caps lock key on, permanently.
- o When entering dates, use the forward slash, (/), between the months, days and years, i.e., 07/28/88.

- o After typing Y or N (Yes or No), hit return.

Options D and E

The main purpose of this module is achieved by reports D and E. If in options D or E, you select Y (Yes), to send the report to the file, then the report will scroll by on the screen after which the main menu comes back, then:

1. Hit G to exit.
2. At the C> type: [LIST LOG.TXT], hit return.
3. Then you will be able to scroll through the report on the screen, backwards and forwards, a line or page at a time and left or right using the left, right, up and down arrow keys, or the page up and page down keys.

If in options D or E, you select N (No), the report will be printed out.

Options A, B, and F

Options A, B, and F are similar in that they all handle the same data. A is for inputting that data, B is for editing it and F is for printing out all inputted data.

Options C and G

Option C is for deleting all data for a specific equipment item.

Option G is for exiting the program back to the C>.

CUSTOMIZED MODIFICATIONS AND CONSULTING SUPPORT

Customized modifications, additions and consulting support to this program are available by contacting Frank Herbaty and Associates at 708-759-1915.

INSTRUCTIONS FOR USING DEMO DISK

Symbol and Notes

- o R means to hit the return (or enter) key.
- o Only those characters enclosed in brackets [ ] are meant to be typed.
- o Spaces shown between characters in brackets are mandatory.

Create a New Directory

1. At the C>, type [MD\DEMO], R
2. Type [CD\DEMO], R

This will put you into the new directory (DEMO) that you have just created.

Copy All Demo Disks into the New Directory

1. Put any of the three disks in drive A (the disks are marked 1, 2 and 3 in red in the lower right hand corner).
2. At the C>, type [COPY A:*.*], R.
3. Repeat step 2 for the other disks.

This will copy all 3 disks into the demo directory on your hard disk.

User's Guide

The User's Guide can be printed out by:

1. Make certain you are in the "demo" directory and have a C>.
2. Press the CTRL key and then tap the letter P.
3. Type [TYPE GUIDE2] at the C>, R.

Running the Program

The User's Guide2 gives full instructions on how to use this software, however, one small change must be made. Refer to page 2 of the User's Guide2. Substitute "DEMO" for "MAINT" for the demo disks. Type [DEMO], R at the C> in the demo directory to start the program. Since the program is menu driven, you really don't need the User's Guide. It's so easy to use.

The demo disk works exactly the same as the regular program, except that it will allow you to enter only 10 records for each of the six programs. However you can add, delete or modify records and print any of the reports and do anything that the actual program can do. If you attempt to add the eleventh record, the entire program you are working in will be incapacitated, therefore you must keep track of the numbers of records in the program and delete one record if you are in danger of exceeding the limit. The way you can see how many records there are is by printing out the following options:

Option M in Program 1	Option H in Program 4
Option H in Program 2	Option C in Program 5
Option E in Program 3	Option F in Program 6

If you do incapacitate any program, it can be reactivated by recopying the disks on the the hard disk by repeating step B above. Put write protect tabs on the original demo disks to prevent anyone from entering data on these disks. If someone does, and enters more than 10 records, the disks will be irretrievably ruined.

The caps lock should be on at all times whenever using this program.

One other point to be aware of is that options E, F, and G under program 3 (Equipment Records) are meant to be printed on wide paper. If you use 8 1/2 x 11 paper you can get everything on by using condensed print (15 cps). Furthermore, option E in program 5 requires both wide paper and condensed print capability.

Bibliography

Cooling, W.C., *Low Cost Maintenance Control*, New York: American Management Association (1973)

Corder, A.S., *Maintenance Management Techniques*, New York: McGraw-Hill Inc. (1976)

Godfrey, Robert S., *MEANS Site Work Cost Data 1982*, Massachusetts: Means, Robert Snow Co. Inc.

Gradon, F., *Maintenance Engineering Organization*, New York: John Wiley & Sons Inc. (1973)

Heintzelman, J., *Complete Handbook of Maintenance Management*, New Jersey: Prentice Hall Inc. (1976)

Higgins, L.R. and Morrow, L.C., *Maintenance Engineering Handbook*, 3rd Edition, New York: McGraw-Hill Inc. (1977)

Hildebrand, James K., *Maintenance Turns to the computer*, Massachusetts: CBI Publishing Co. Inc. (1972)

Husband, T.M., *Maintenance Management and Terotechnology*, England: Saxon House (1976)

Kelly, Anthony and Harris, M.J., *Management of Industrial Maintenance*, Massachusetts: Butterworth (1978)

Lewis, Bernard T., *Developing Maintenance Time Standards*, New York: Farnsworth Publishing Co. Inc. (1967)

Lewis, Bernard T., *Management Handbook for Plant Engineers*, New York: McGraw-Hill Inc. (1977)

Lewis, Bernard T. and Tow, L.M., *Readings in Maintenance Management*, Massachusetts: CPI Publishing Co. Inc. (1973)

Mann, Lawrence, *Maintenance Management*, Massachusetts: Lexington Books (1976, 1983)

Morrow, L.C., *Maintenance Engineering Handbook*, New York: McGraw-Hill Inc. (1966)

Newborough, E.T., *Effective Maintenance Management*, New York: McGraw-Hill Inc. (1967)

Ouchi, William, *Theory Z: How American Business Can Meet the Japanese Challenge*, New York: Avon Books (1982)

Patton, Joseph D., Jr., *Maintainability and Maintenance Management*, North Carolina: Instrument Society of America (1981)